THE
BIOLOGY

Activities Book

VIC CHOW

City College of San Francisco

Illustrations by
Karen Chow

KENDALL/HUNT PUBLISHING COMPANY
2460 Kerper Boulevard   P.O. Box 539   Dubuque, Iowa 52004-0539

Contents

Preface

This book was written for those who wish to learn biology in a non-traditional fashion; the biology coloring books on the market are excellent examples of non-traditional approaches. By no means should this book be substituted for a biology textbook, nor was that the intention, but rather it should be used as an extension for learning subject matter in a variety ways.

The purpose of *The Biology Activities Book* is:

1. To reinforce biological concepts learned through activity oriented games or tasks.
2. To develop a sense of critical thinking and analytical thinking not only in the biological sciences but also as a way of dealing with problems in our daily lives.
3. To demonstrate that biology does not always deal with the analytical, scientific, logical, and mathematical approaches but is also artistic, innovative, imaginative, and creative.
4. To allow a diversity of people to enjoy this book regardless of their background in biology. Whether one is a biology teacher, a biology major, or a layperson each should be able to derive some amount of enjoyment and satisfaction with the activities presented in this book.

Let's begin with an example given below entitled, ''I've got your number.''

I've Got Your Number

1. Pick an odd number between 5–55

   Multiply the number times two (2) _____

   Add one (1) to the number _____

   _____

2. Pick another odd number (a different one!) between 5–55 _____

   Multiply the number times two (2)

   Add one (1) to the number _____

3. Subtract the smaller number from the larger number   _____

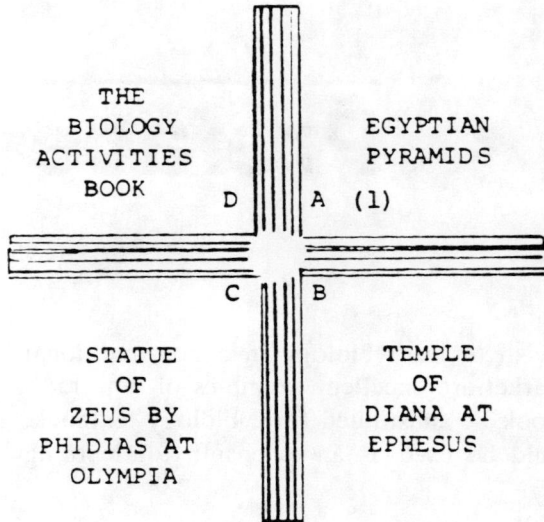

THE
BIOLOGY
ACTIVITIES
BOOK

EGYPTIAN
PYRAMIDS

D   A (1)

C   B

STATUE
OF
ZEUS BY
PHIDIAS AT
OLYMPIA

TEMPLE
OF
DIANA AT
EPHESUS

Starting with the letter A, count in a clockwise direction, to letter B, letter C, letter D and back to letter A the number you had obtained in part 3 above.

Surprise! You have landed on . . .

Answer: *The Biology Activities Book*

It is not to say that *The Biology Activities Book* is to be compared with some of the Seven Wonders of the World but it seems that no matter what you do, you'll always end up with it. Have fun embarking on a strange but rather enlightening approach to learning biology.

Contents of the Activities Book

The activities book was arbitrarily separated into 9 parts covering topics from botany, ecology, human anatomy to zoology. Ah yes, a little of chemistry is provided as well. It was my intention to encompass as many different areas as possible to give readers an array of activities to enjoy.

Some of the activities will require the use of color pencils or pens, a calculator, scissors, and glue. Generally the minimum required is a pen or a pencil and some patience. It is highly recommended that one has a reference text such as a biology textbook, a dictionary, an encyclopedia, etc. in order to check over answers where diagrams or pictures are used. In any case, answers will be provided in the back of the book according to the page number of the activity. In a couple of activities presented the reader will require a partner in order to complete the exercise (but most of the activities can be completed by the individual).

Each of the sections (9) has been separated with a cover page in order to give the reader a more detailed discussion on the contents and usage of the section. Please give your utmost attention to these cover pages.

Acknowledgments

This project started three years ago when a few of my students challenged me to develop a fun book for biology. They had not seen one on the market nor had I. Looking over the magazine racks at the airport one day, I noticed many activities for people on the go such as crossword puzzles, word search books, cryptograms, game magazines, etc. Little by little, I tried to develop something that could be enjoyed by my students. To them I owe a load of thanks. Family of course, Kathy, David, Karen continues to encourage my growth and pleasure in life and the teaching profession.

PART ONE

General Puzzles

A compendium of mixed games, mazes and puzzles illustrating the many approaches one may use when studying, thinking, visualizing, creating, and solving problems. For example, ''A Pretest to General Biology'' attempts to force you to arrive at the correct answers thereby allowing you to learn the subject matter being taught; ''I Think I Can'' gives you a chance to use something that is probably familiar to you but you need to be somewhat creative and imaginative to arrive at an answer; and ''Cryptic Biology—The DNA Story'' is a game of codes—each cartoon is a complete message of its own when correctly solved.

A Pretest to General Biology

C	O	M	P	U	T	E	R	I	Z	E	D	M	E	S	S	A	G	E
1	2	3	4	5	6	7	8	9	10	11	12	13	14	15	16	17	18	19

1. If the human heart has four chambers change the C in square 1 to W; if not, change the C in square 1 to S.

2. If the functional unit of life is the tissue change the S in square 15 to U; if not change the S in square 15 to L.

3. If diffusion occurs in only the gaseous state change the letter T in square 6 to E; if not change the letter T in square 6 to M.

4. If a testes is an example of a gonad change the letter D in square 12 to B; if not change the letter D in square 6 to M.

5. If the coverings of the human brain is called meninges change the letter G in square 18 to Y; if not change the letter G in square 18 to P.

6. If the large intestine digests and absorbs all foods change the letter E in square 11 to U; if not leave the letter E in square 11 blank.

7. If an example of a complex carbohydrate is glucose change the letter P in square 4 to K; if not change the letter P in square 4 to C.

8. If cells come from pre-existing cells leave the letter in square 7 unchanged; if not change the letter in square 7 to A.

9. If Mendel is the Father of Heredity change the letter O in square 2 to E; if not leave the letter O in square 2 unchanged.

10. If a person of blood type A can donate blood to a person of blood type B change the letter I in square 9 to K; if not, change the letter I in square 9 to T.

11. If an example of a long bone is the femur change the letter A in square 17 to G; if not change the letter A in square 17 to K.

12. If the symbol for calcium is "C" change the letter U in square 5 to R; if not change the letter U in square 5 to O.

13. If cardiac muscle is an example of voluntary muscle change the letter S in square 14 to L; if not change the letter S in square 14 to O.

14. If an example of tissue is epithelium change the letter S in square 16 to O; if not change the letter S in square 16 to X.

15. If DNA contains the base uracil change the letter M in square 13 to the letter I; in any case change the letter M in square 13 to the letter I.

16. If another word for trachea is windpipe change the letter M in square 3 to L; if not change the letter M in square 3 to R.

3

17. If the functional unit of the kidneys is the neuron change the letter R in square 8 to S; if not leave the letter R in square 8 blank.

18. If normal blood pressure for young adults is 150/90 mm change the letter Z in square 10 to Q; if not change the letter Z in square 10 to O.

Bonus—change the letter E in square 19 to an exclamation mark!

1	2	3	4	5	6	7	8	9	10	11	12	13	14	15	16	17	18	19

Walled-In

A lymphocyte is required to cleanse each of the columnar cells by methodically moving through every cell membrane. Starting at point A, show the pathway that the lymphocyte may take in order to complete its task. Oh yes, there is a restriction—the lymphocyte must move in one continuous pathway (line) and it may only pass through each cell membrane *once*.

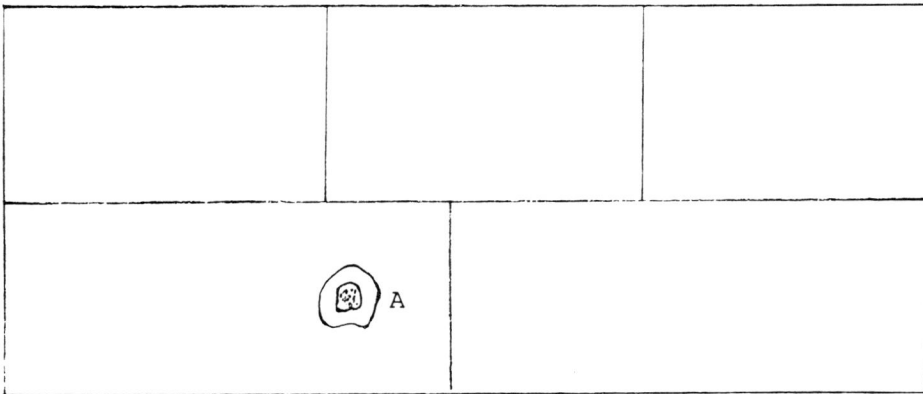

Confused Anatomy

A poor student has learned his anatomy incorrectly and has arranged the organs in the following order:

H	K	S	B	N
E	I	T	R	O
A	D	O	A	S
R	N	M	I	E
T	E	A	N	
	Y	C		
		H		

Place the organs in their correct sequence given the following pieces of information:

The BRAIN is next to the KIDNEY

The STOMACH is one organ away from the BRAIN

The HEART and KIDNEY are not next to each other

The NOSE is second to the left

The Pluses and Minuses of Life

Help the minus charge (circled) leave the puzzle at the *end* site. You must always move alternately from a minus to plus and a plus to minus sign in either a vertical or horizontal direction. You may not move diagonally.

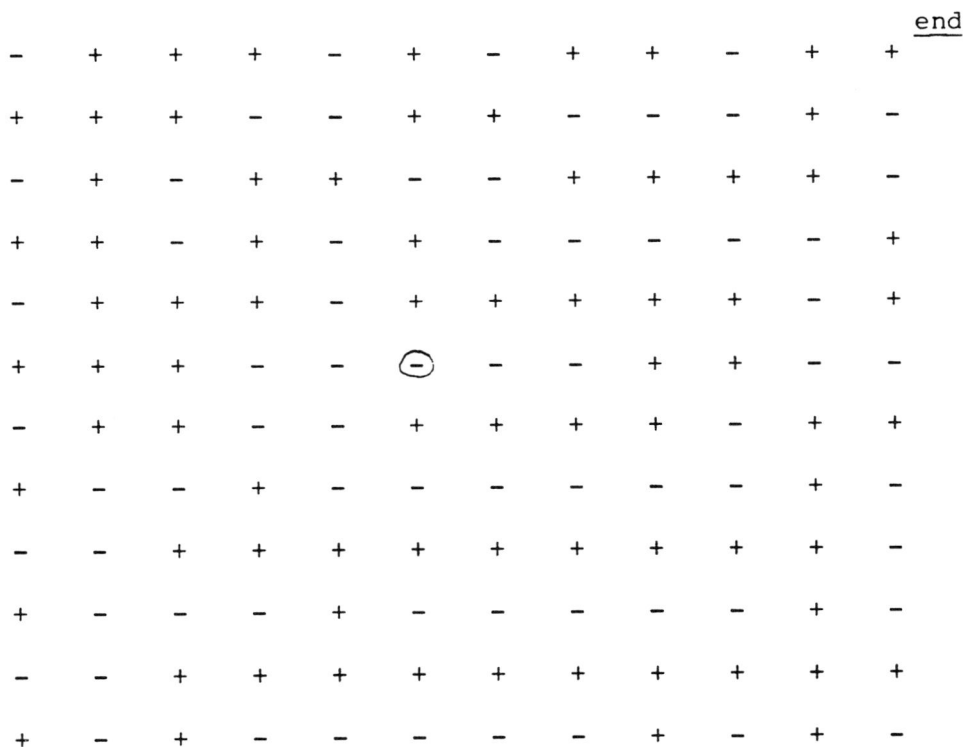

```
                                                              end
  −    +    +    +    −    +    −    +    +    −    +    +

  +    +    +    −    −    +    +    −    −    −    +    −

  −    +    −    +    +    −    −    +    +    +    +    −

  +    +    −    +    −    +    −    −    −    −    −    +

  −    +    +    +    −    +    +    +    +    +    −    +

  +    +    +    −    −   (−)   −    −    +    +    −    −

  −    +    +    −    −    +    +    +    +    −    +    +

  +    −    −    +    −    −    −    −    −    −    +    −

  −    −    +    +    +    +    +    +    +    +    +    −

  +    −    −    −    +    −    −    −    −    −    +    −

  −    −    +    +    +    +    +    +    +    +    +    +

  +    −    +    −    −    −    −    −    +    −    +    −
```

I Think I Can

A very special game called THINK was developed to allow you to . . . what else . . . think. It is played very much like BINGO where the letters and numbers are called out. The difference is that in the game of THINK, letters are also placed in the squares to reveal a hidden message. Below are a series of letters and numbers for you to decipher. Alas, the printer forgot to include the numbers! Now with some clues given (T-1 = E, T-2 = V, T-15 = O, H-26 = Q, H-28 = R, K-61 = G, K-63 = E) think about where the numbers should go in order for you to decode the entire message.

THINK

E	C	O	M	G
R	A	E	S	N
T	E	N	I	B
A	!	D	R	F
N	R	O	C	A

THINK

V	A	H	B	C
O	D	I	I	N
N	T	U	X	L
A	Q	O	I	R
N	E	S	S	U

THINK

L	E	O	U	E
E	U	N	T	T
R	S	I	C	E
A	N	C	E	R
O	E	N	Y	E

Message

T-1, H-26, N-48, K-73, T-3/ H-17, N-46, T-15, I-38, I-37, K-66, H-24/ T-5, K-70/N-54, H-19, K-68, I-33, H-28, N-50, N-57, N-59/ K-62, T-10, T-14/ N-47, T-6/ K-67, H-21, N-55, I-36, K-75, I-40/ H-20, K-74, T-4, I-39, T-8, K-61/ N-56, K-64, T-7, H-29, H-27, N-49, K-69/ H-16, I-41, K-65, N-58, K-63, T-13, H-23, T-9, T-11, N-51, I-35, I-31, I-45/ I-43, K-72/ I-32, I-34, T-12, T-2, N-60/ H-22, N-53, H-18, K-71, I-42, N-52, I-44, H-30, H-25/

It's Your Move!

Instructions: In eight moves, go from start to finish and pass through each of the characteristics of living things once.

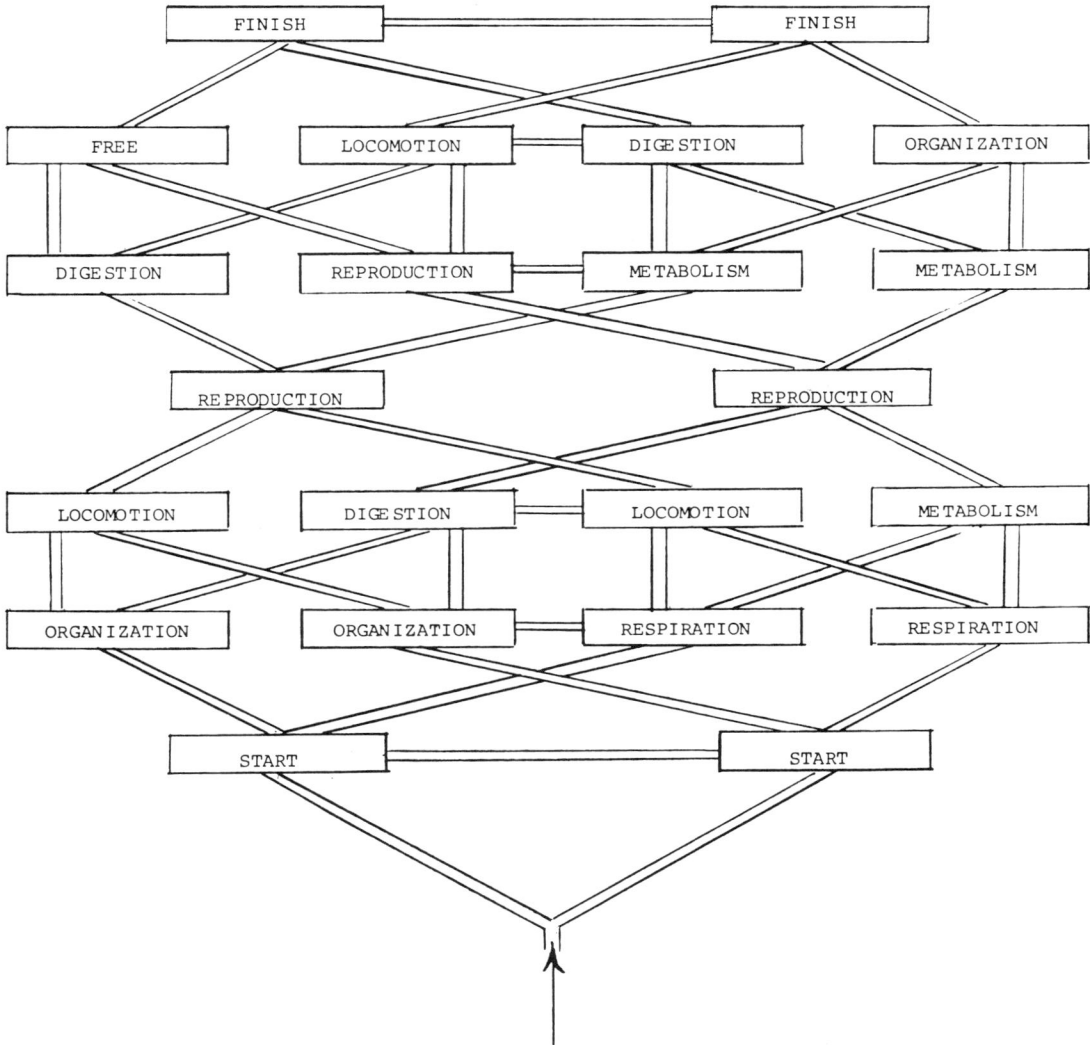

Indigestion

Help the organism digest his fatty foods in the quickest way possible by allowing the fats to alternate between the B (bile) blocks and the L (liver) blocks either vertically or horizontally but never diagonally starting at the oral cavity and ending at the funnel cavity.

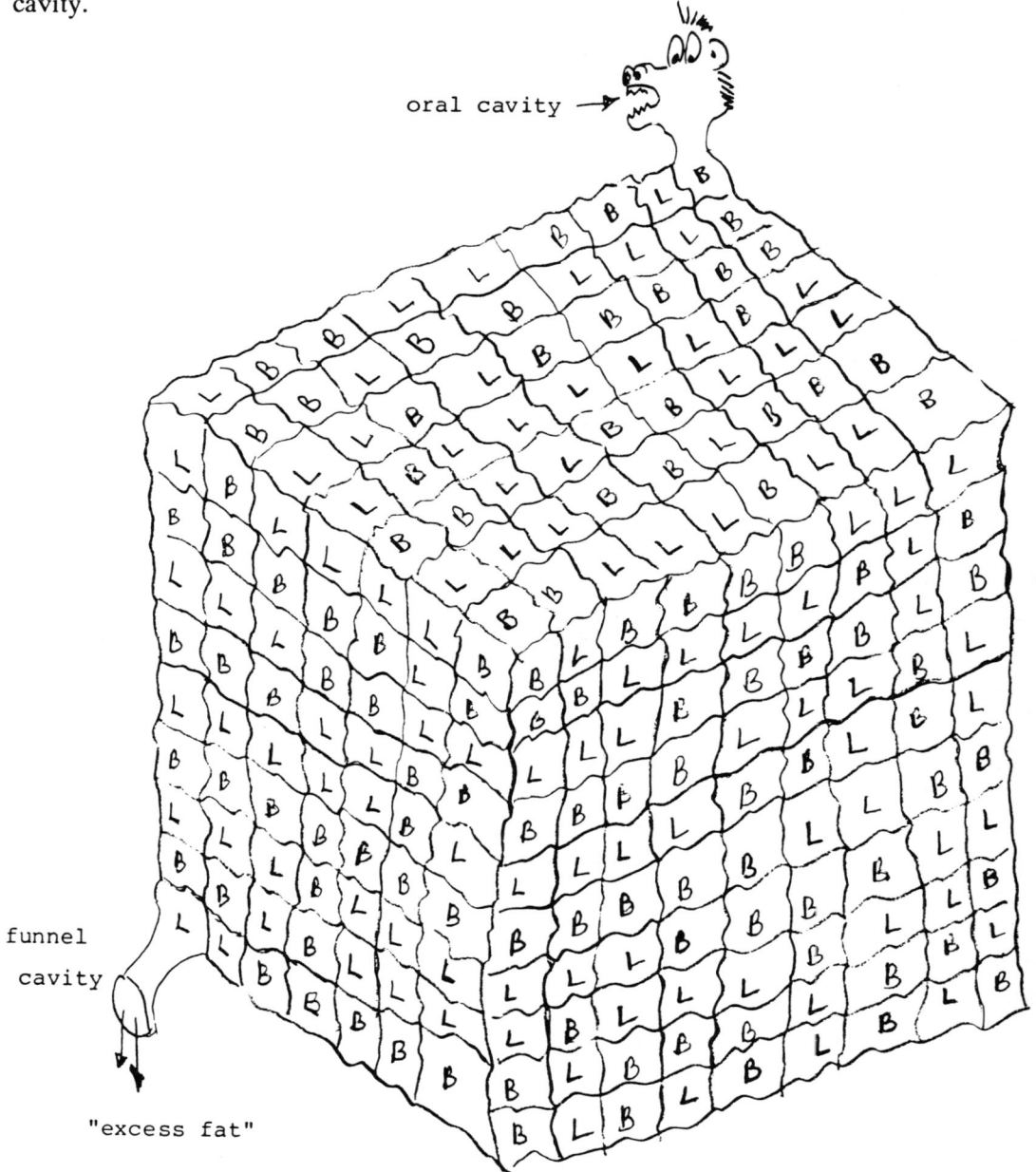

oral cavity →

funnel cavity

"excess fat"

10

Paradise (Pairing Dice)

A die containing names of tissues is shown below in three different positions. Pair up the tissue by determining the names of tissues on opposite sides of the die. One of the six tissues is not shown in the diagram; it is a five letter word related to the saying, "_____, sweat, and tears."

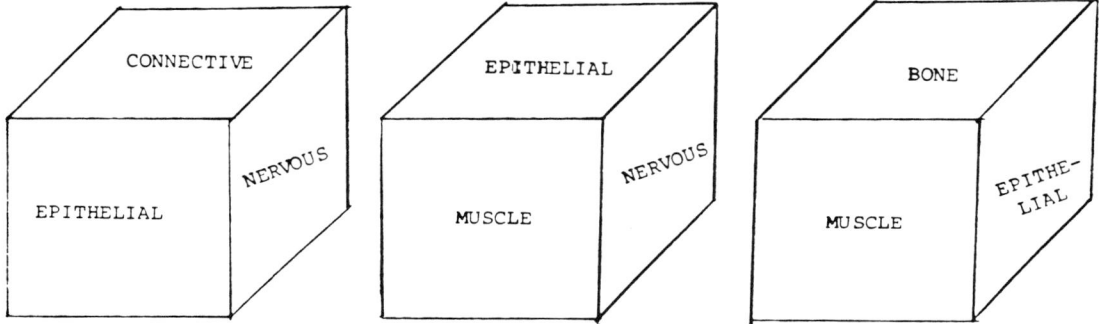

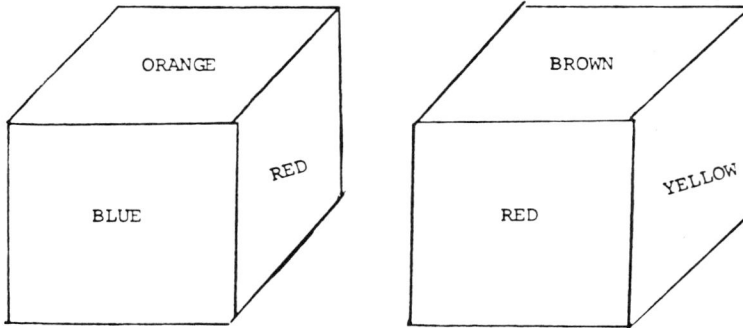

How about pairing colors given two views?

Cryptic Biology—The DNA Story

Part I

Clue: The letters represented below fall one short of the real letters in our story . . . break the code.

Cryptic Biology—The DNA Story

Part II

Clue: The letters represented below fall two short of the real letters in our story . . . break the code.

Cryptic Biology—The DNA Story

Part III

Clue: The first three words in one of the cartoon panels of Part I using the new code would be, "CEE NOJF BCGHJF" . . . break the code.

DNA Round-up

The DNAs are on the loose! Draw three cell membranes (straight lines) so that each DNA is isolated from each other. Also, try the wordy problems below.

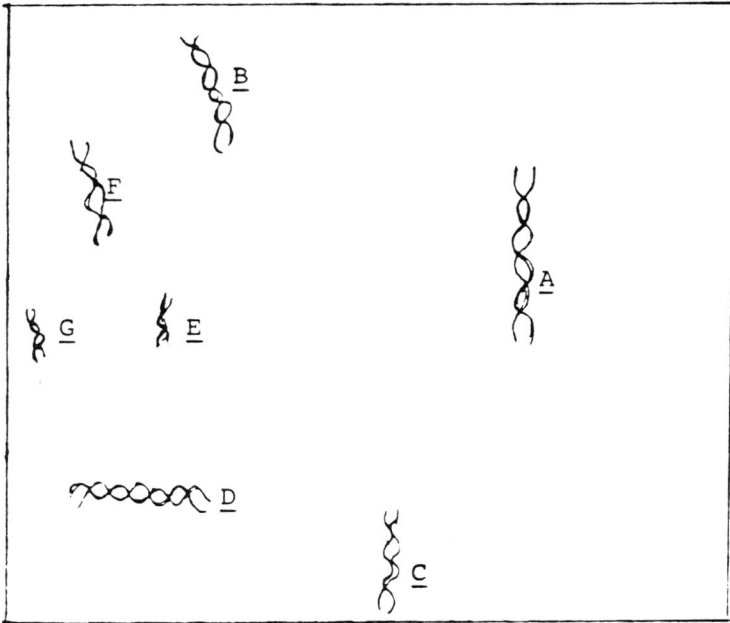

Wordy Problem

1. DNA *G* is composed of a (nucleotide) strand that has twice as many purines as there are pyrimidines, where the number of purines are said to be equal and that the number of thymines is half the number of cytosines. The total number of bases in the strand is 27.

2. DNA *A* has a (nucleotide) strand that is complementary to the DNA described in DNA *G* except that it has an additional number of adenines that is equal to the number (of adenines) present.

3. DNA *F* is composed of a (nucleotide) strand that contains two-thirds the number and types of bases found on the strand of DNA *A*.

4. DNA *B* is composed of a (nucleotide) strand of DNA that contains two more of each purines that is found on the strand of DNA *F*.

5. Finally . . . construct both strands of DNA *C* given the fact that one strand (nucleotide) resembles DNA *B* strand but with two less of each of the pyrimidines.

And what about DNA *D*? It's taking a siesta and therefore could not be accounted for. . . .

What a Move!

It is said that it takes only a minute or so for blood to travel through all the blood vessels of the body. How is this done? Actually the secret is found in the presence of "tiny valves" found in the veins of the circulatory system. Below is such an example. See if you can help the little rbc (red blood cell) move from its starting point to its final destination by moving through the valves in the correct direction; be careful, some of the valves are faced in the wrong direction and may cause a blood clot (one has already formed!).

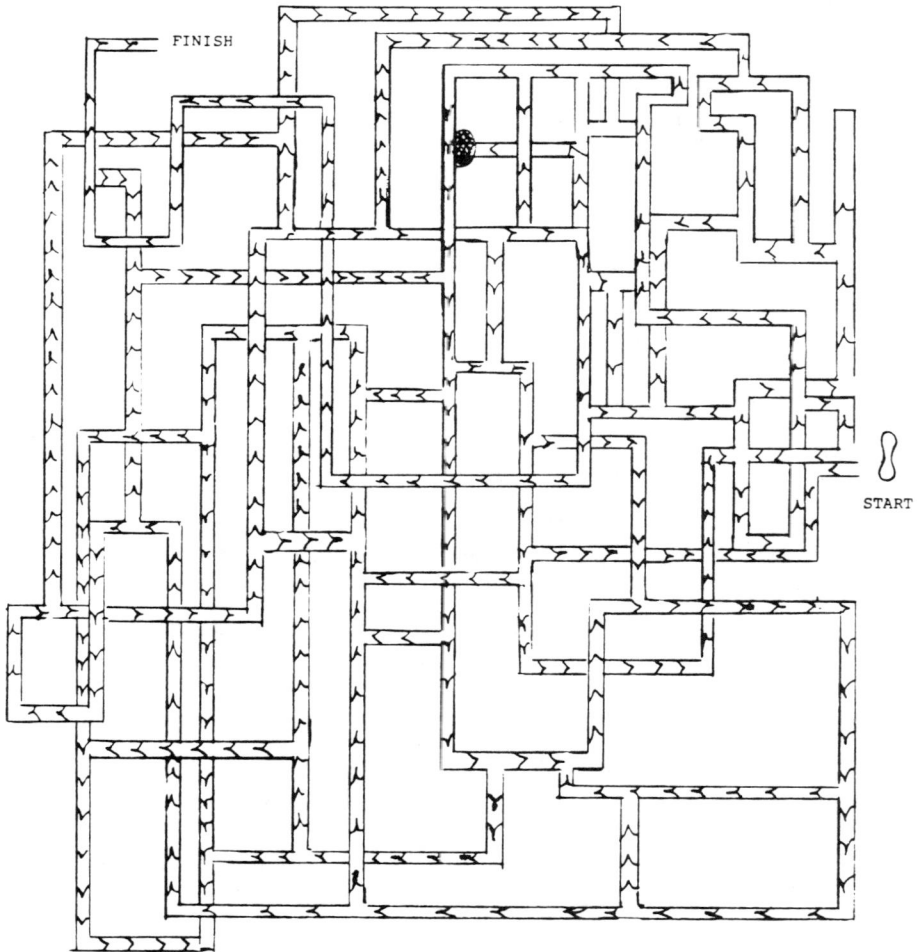

FINISH

START

PART TWO

Word Games

A series of biological words or terms are introduced as being mixed, scrambled, misspelled, related or hidden that requires one to make the necessary corrections to solve the problem. These games test your familiarity with common words or terms used in the biological sciences.

Unscramble

The following words have completely scrambled among one another. They either have the wrong prefix, root word, or suffix. What are the real words?

Biological Terms

1. Herbderm

2. Antipermeable

3. Chloroscope

4. Histzoan

5. Hemogram

6. Protometer

7. Cardiophyll

8. Centiology

9. Micrology

10. Epiglobin

11. Anthropaceous

12. Protology

13. Semiseptic

14. Bioplasm

15. Diffusion

Biological Phenomenon

Instructions: The last two letters of the first word is the beginning two letters of the second word. Fill in the missing letters.

STI___RITHECIUM

COLUMN___EOLAR

AC___URON

STIG___TING TYPE

MERIST___BRYO

DIALYS___OTONIC

CHROMAT___TERPHASE

PROTE___ORGANIC

RIBOSO___IOSIS

ENZY___GAVITAMINS

JAUNDI___CUM

RUM___DOPARASITE

STOMA___YME

PACEMAK___YTHROCYTES

FLOW___OSION

STAR___LOROPLASTS

PLAS___CROPHAGES

OXYHEMOGLOB___TERCOSTALS

ALDOSTERO___PHRON

URI___PHRON

ANTIG___DOTOXIN

GENERATI___TOGENY

PHENOTY___RFECT FLOWER

PHYL___BEL

ASEXU___TERNATION OF GENERATION

CAPSU___NTICELS

ISOGA___COLOGY

CUTIC___AF

EPIPHY___NDRIL

FUNGICI___RMATOPHYTES

CHLOROS___OGAMY

20

Reproductive Errors

Instructions: Remove the extra letter or letters from the misspelled words and place them in sequential order below to find a surprise message.

1. CHHLOROPHYLL
2. PROTEEIN
3. AATOM
4. NERVVE
5. HEREDITYY
6. DEGRADDATION
7. EUCARRYOTE
8. FUNGII
9. NNEBULAR
10. QUARKK
11. METABOLIISM
12. ENNERGGY
13. COSMMOS
14. MEEMBRANNE
15. CAARRBON
16. ELEEMENT
17. NUCLLEIC ACIIDS

18. CYTOKKINEESIS
19. ELLECTRONS
20. OXYYGEN
21. METTAPHASE
22. RIBOSOOME
23. MEIOSSIIS
24. CHRROMOSOMEE
25. PHAGGOCYTOSIIS
26. PERRMEABLLE
27. OSSMOSIS
28. CUBBOIDAL
29. PARENCHYYMA
30. CORTTEEX
31. COLUMNNAR
32. MYOFFILAMENT
33. MESOOPHYLLL
34. EPIDDERMISS

H _ _ _ _   _ _ _ _ _ _ _   _ _ _   _ _ _

_ _ _ _ _   _ _   _ _ _ _   _ _ _ _ _   _ _

_ _ _   _ _ _ _ _ !

Words: An Anagram

Rearrange the letters to spell out common words in biology and decide which of the words (underline one) in each set doesn't belong with the others. Also, give a reason for your answer below.

1. THARES   LOCNO   PELESN   CHAMOTS

2. ROBCAN   RION   OSLEUGC   GEXONY

3. NOCR   SPEA   ALECER   CHINAPS

4. UDIOMS   LRIHOCED   DERILUFO   IDEOID

5. ELAGA   EABAMO   UNFIG   SOMS

6. WOC   PESHE   RIGTE   DERE

7. ORPTEWAM   CKTI   CHELE   LEAF

8. ALLECLWL   (two   words)   TECIORENL   TPHLOCROASL
   HLYLLORCOHP

9. CDACURII (two words)   AMAMNOI   AREU   IREPTON

10. LEE   SFHI   RAKHS   NANMEEO

Reasons:

1.

2.

3.

4.

5.

6.

7.

8.

9.

10.

Hide and Seek

The digestive system is always causing problems hiding its structures among the letters below. Find and circle those structures listed below (twelve in all). Work vertically, horizontally or diagonally but always in a straight line. Oops, did I mention that some of the words are spelled backwards?

```
P  H  C  A  M  O  T  S  O  R  A  L
C  A  V  I  T  Y  O  C  A  B  L  N
E  S  U  G  A  H  P  O  S  E  I  E
R  U  L  R  E  V  I  L  E  G  M  E
R  E  D  D  A  L  B  O  N  A  E  L
B  E  U  G  N  O  T  N  O  L  N  P
I  N  T  E  S  T  I  N  E  L  T  S
```

Words List

STOMACH	ESOPHAGUS	COLON
SPLEEN	LIVER	GALL
CAVITY	BLADDER	INTESTINE
ORAL	ALIMENT	TONGUE

Dear Heart

The message is dear to my heart after you remove eleven words. What is the message?

```
A   N   E   O   T   H   E   L   R   W

O   R   D   E   F   O   R   V   S   E

I   N   O   A   N   T   R   I   W   A

L   N   O   O   D   E   I   S   R   P

A   D   C   E   M   A   S   K   E   R
```

Word Search

Hold It . . . I Think You're Going to Like This—Rearrange the order of the *columns* of letters to find hidden words that are related to the excretory system. There are thirteen words in all. You may wish to place the new columns of letters in the checkerboard provided for you on the page below.

#1	#2	#3	#4	#5	#6	#7	#8
N	E	S	K	I	B	D	Y
K	I	U	P	E	L	S	N
A	T	E	S	W	A	E	O
A	P	C	U	R	D	E	U
O	N	A	A	M	D	M	I
V	E	I	K	L	E	I	R
I	E	R	C	A	R	R	D
T	E	B	E	W	O	A	R
C	A	Y	R	T	X	E	L
C	A	I	U	R	D	I	C
Y	G	N	U	O	T	X	E

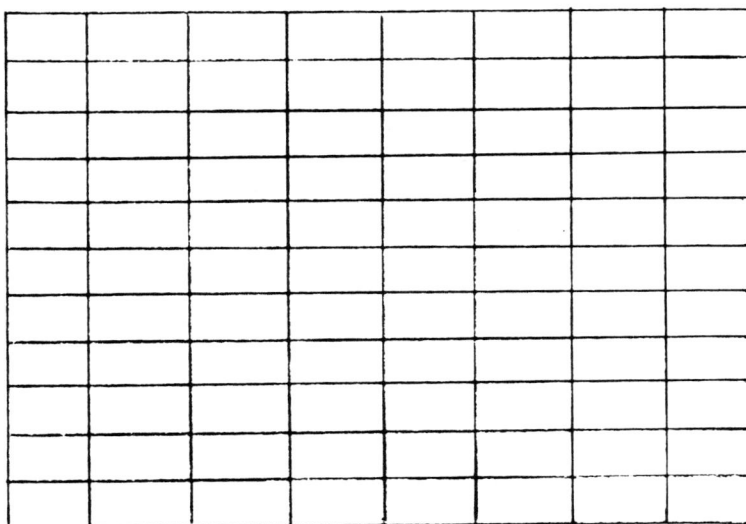

25

And the Word Is . . .

Instructions: Below are squares that are partially completed. Your task is to fill in the blanks (to complete the word) given two sources of information: (a) the clue which is given immediately besides each set of squares and (b) the arrows which indicate which letters are to be carried over to the next word. The first word has been done for you as an example. Oh yes, at no extra cost to you, I've thrown in a few extra letters in the squares to make it worth your while.

0. What this book is all about

1. Single cell animal

2. Plant organelle

3. Powerhouse

4. Chemical activity

5. Hereditary material

6. Organisms with backbones

7. Also known as sinoatrial node

8. A leech is an example of this

9. Cone bearing plants belong to this class

10. Neural-endocrine site

11. Loss of electrons from an atom or molecule

12. A term to end or to complete

| B | I | O | L | O | G | Y |

| | | O | | O | Z | | | | |

| | H | | O | | | | | S | |

| | I | | | | | D | | | |

| | E | | | | | | M | |

| | H | | | | M | | | |

| | | T | | | | | E | |

| | | | M | | | | | |

| | T | | A | | | I | | | |

| | | N | S | | | | | | |

| | | | | H | A | | | u | |

| | | D | | | O | | | |

| | | | | | | |

Mixed Arrangements

Part I

Instructions: Rearrange the order of the letters in each set of letters to spell biological terms. For example: TIM OCH OND AIR when rearranged will spell out the word MITOCHONDRIA. Watch out—some words are singular and some words are plural.

1.	IOB	GLO	TIS	
2.	SIT	ESU		
3.	RCA	HOB	DRY	EAT
4.	HEC	SMI	TRY	
5.	EEL	RCT	SON	
6.	PIL	SID		
7.	LOM	CUE	SEL	
8.	ANT	CIO	NOD	
9.	POE	NOR		
10.	NEC	RIT	LOE	
11.	HRC	AMO	NIT	
12.	CIM	ROT	BUU	LES
13.	BIR	SOO	SEM	
14.	REP	EAM	LEB	
15.	CIM	ROT	MOE	
16.	FID	SUF	ION	
17.	CAR	BUN	CLE	
18.	CED	DIU	SOU	
19.	PIE	RED	SIM	
20.	CES	TER	NIO	

Mixed Arrangements

Part II

Instructions: Rearrange the order of the letters in each set of letters *and* the order of the sets of letters to spell biological terms. For example: AIR TIM OND OCH when rearranged (twice) will spell out the word MITOCHONDRIA. Watch out—some words are singular and some words are plural.

1. OPH   OTR   AUT
2. AIN   NIH   DBR
3. ACT   NIO   FOL
4. NIE   HYT   XOR
5. LIT   RAC   AGE
6. TYE   COE   TOS
7. ONE   LAG   STL
8. DEE   SET   TAN
9. ION   GIN   TES
10. MIA   CYY   HET   LOP
11. TZA   INU   MIM   NIO
12. COL   RAG   UNA   TYE
13. ISS   CER   BUT   LOU
14. POE   ROO   SIS   SOT
15. RAY   ILL   PAC
16. MAE   PEM   SHY
17. TIE   END   SAR   POA
18. CHC   SED   RIA   ASO   MNO
19. BET   ARE   VRE
20. ROE   THA   ORE   ISS   CLS

Mixed Arrangements

Part III

Instructions: Rearrange the order of the letters in each set of letters to spell biological terms related to botany. For example: ACC SUT when rearranged will spell out the word CACTUS. Watch out—some words are singular and some words are plural.

1. SAT   NEM
2. NUR   REN
3. LHP   MOE
4. TAN   REH
5. UMS   HOR   SOM
6. HOR   EST   LIA
7. VIL   REW   ROT
8. ECD   IUD   USO
9. PIE   HYP   SET
10. NAN   LUA   AIR
11. CEH   NST   SUT
12. LCH   OOR   ISS
13. REX   POH   TEY
14. SME   OHP   LYL
15. REP   CIY   CEL
16. SIP   ROG   ARY
17. BIG   BRE   LEL   SIN
18. MAG   TOE   YHP   SET
19. GAN   SOI   REP   MAE
20. HLC   ORO   LAP   TSS
21. NOC   IID   SOP   ROE
22. YHP   OCM   CEY   SET
23. HAT   LOL   YPH   EST
24. GEM   GAA   TEM   PHO   EYT
25. OHP   POT   RIE   DIO   SSM

29

Mixed Arrangements

Part IV

Instructions: Rearrange the order of the letters in each set of letters to spell biological terms related to ecology. For example: NEE GRY when rearranged will spell out the word ENERGY. Watch out—some words are singular and some words are plural.

1. CIN   SHE
2. IOB   CIT
3. ORG   TWH
4. TAW   SER   DEH
5. ROP   UCD   SER
6. RAP   IAS   SET
7. OCE   SSY   MET
8. NCO   MUS   RSE
9. ESR   ROU   SEC
10. REP   TAD   NOI
11. RHE   BIV   ROE
12. MCO   NMU   YIT
13. OBI   HPS   REE
14. RCA   VIN   ORE
15. TAU   ROT   POH
16. IVD   SRE   IYT
17. VEO   TLU   NOI
18. TEH   REO   ROT   SPH
19. TED   TIR   VIO   SRE
20. PEX   IOL   ATT   ION
21. MCO   ENM   ALS   MIS
22. TUM   LUA   SMI

Mixed Arrangements

Part V

Instructions: Rearrange the order of the letters in each set of letters to spell biological terms related to zoology. For example: AIN LAM when rearranged will spell out the word ANIMAL. Watch out—some words are singular and some words are plural.

1.	HIC	NOT		
2.	BOE	AIL		
3.	ECO	MOL		
4.	POS	GEN		
5.	SIN	TEC		
6.	IFS	SHE		
7.	SAN	ESK		
8.	ZLI	RAD		
9.	AMM	LAM		
10.	TEM	ZOA	SAN	
11.	HIC	ARD	NIA	
12.	FAL	TWO	SRM	
13.	MEN	TOA	SED	
14.	TOC	OUP	SSE	
15.	RAT	ORH	OPD	
16.	MAP	BIH	NAI	
17.	COH	RAD	TES	
18.	TER	TAM	OED	
19.	MEN	TER	ANE	
20.	LIB	TEA	LAR	
21.	GAS	ROT	ODP	
22.	GSE	MNE	ATT	ION
23.	NVI	TER	REB	ATE
24.	EOX	SEK	TLE	SON
25.	CEH	CLI	ERA	ETS
26.	ROU	HCO	ADR	SET

How's Your System Functioning?

Part I

Instructions: In each of the squares below are letters that spell out various organs related to a specific organ system. By combining the letters horizontally from the various rows identify the organ system and the organs.

ST	GE	CR	AC	NE
L	AN	M	TI	AS
P	TE	V	TI	R
IN	I	S	E	VE
DI	O	S	E	H

ORGAN SYSTEM _____

ORGANS

_____   _____

_____   _____

UR	EP	DN	R	N
B	IN	H	RO	S
UR	LA	A	E	YS
K	E	DD	R	Y
N	I	TE	E	R

ORGAN SYSTEM _____

ORGANS

_____   _____

_____   _____

How's Your System Functioning?

Part II

Instructions: In each of the squares below are letters that spell out various organs related to a specific organ system. By combining the letters horizontally from the various rows identify the organ system and the organs.

PR	CR	IDY	T	US
T	OS	DUC	M	UM
REP	E	TA	TE	S
EP	RO	O	TI	E
S	ID	S	T	VE

ORGAN SYSTEM _____

ORGANS

_____   _____

_____   _____

T	N	UI	OI	NE
PA	HY	EN	RI	LS
A	IT	C	A	RY
EN	DR	R	TA	AS
P	DO	C	RE	D

ORGAN SYSTEM _____

ORGANS

_____   _____

_____   _____

33

Instructions: In each of the squares below are letters that spell out various organs related to a specific organ system. By combining the letters horizontally from the various rows identify the organ system and the organs.

HY	ER	U	BR	MUS
ME	R	THA	LL	US
N	PO	U	RO	UM
CE	D	V	LA	N
N	E	E	O	A

ORGAN SYSTEM _____

ORGANS

_____   _____

_____   _____

O	ST	L	RP	US
ME	E	U	ET	N
S	A	E	O	AL
R	K	CA	L	ALS
SK	TA	D	I	L

ORGAN SYSTEM _____

ORGANS

_____   _____

_____   _____

Clue

Instructions: Below are statements with regards to biological structures or processes. Read over the clues in column I and write your answers in the blank provided (on the far right side of the paper). Column I clues will also direct you to Column II clues in the event you are unable to arrive at an answer (letters in parentheses). Finally, you are directed to use your set of answers against a word search puzzle provided for you on the following page. If all else fails, check the answers in the back of the book.

Column I	**Column II**	**Answers**
1. A type of metabolism (a)	ad. Chromosome line at the equator.	1. _____
2. An animal organelle (b)	t. ATP production	2. _____
3. A type of cell division (c)	m. Another term for kernel	3. _____
4. Composition of cell membrane (d)	h. Magnifies objects	4. _____
5. A chemical compound (e)	w. Site of protein synthesis	5. _____
6. A plant organelle (f)	aa. Starch-like carbohydrate	6. _____
7. An inorganic molecule (g)	z. Storage place	7. _____
8. A scientific instrument (h)	u. Also known as ER	8. _____
9. Interdependent cell units (i)	o. Lysosomes perform this function	9. _____
10. Boundaries of a cell (j)	j. Membranous	10. _____
11. Living substances within a cell (k)	v. Packaging site	11. _____
12. Non-functioning substances within the cell (l)	a. From simple to complex	12. _____
13. Brains of the cell (m)	x. Hollow tubes	13. _____
14. Substance found within the nucleus (n)	f. Photosynthesis	14. _____
15. A type of metabolism (o)	g. Universal solvent	15. _____
16. A type of nucleic acid (p)	ab. Green molecule	16. _____
17. Substance found within the nucleus (q)	r. Does not contain uracil	17. _____

Column I	Column II	Answers
18. A type of nucleic acid (r)	i. Small organs	18. _____
19. A dissolving body (s)	d. Hydrophobic	19. _____
20. Powerhouse of the cell (t)	s. Digestive enzymes	20. _____
21. Complex passageways of the cell (u)	p. Contains uracil	21. _____
22. Plant/animal organelle (v)	n. Color bodies	22. _____
23. Bodies related to the ER (w)	b. Cell division	23. _____
24. Structures that form cilia (x)	c. Daughter cells are identical	24. _____
25. Structures composed of solid fibers of strong protein (y)	l. Starch granules as an example	25. _____
26. A membranous sac (z)	e. Peptide bonds	26. _____
27. Plant cell membrane (aa)	k. Bounded by the nuclear membrane	27. _____
28. Photosynthetic organelle (ab)	q. Produces new RNA	28. _____
29. Type of cell division (ac)	y. Aids in locomotion or movement	29. _____
30. Stage of cell cycle (ad)	ac. Division of cytoplasm	30. _____

Word Search

The Cell-structure and function. Using the list of words determined on the previous exercise check to see if they appear in this puzzle. Work horizontally, vertically or diagonally but always in a straight line.

```
C  M  I  T  O  C  H  O  N  D  R  I  O  N  T  V  E
H  P  A  C  H  L  O  R  O  P  L  A  S  T  S  O  N
R  K  E  N  C  E  L  L  M  E  M  B  R  A  N  E  D
O  U  R  A  A  N  U  C  L  E  O  L  U  S  U  C  O
M  O  N  T  H  B  L  Y  S  O  S  O  M  E  C  A  P
O  W  A  T  E  R  O  N  O  T  H  I  N  G  L  T  L
S  C  T  A  R  C  E  L  L  U  L  O  S  E  E  A  A
O  V  I  R  E  E  Y  M  I  T  O  S  I  S  O  B  S
M  I  N  D  E  N  A  C  E  S  O  O  N  L  P  O  M
E  V  E  N  O  T  N  I  R  A  M  V  C  G  L  L  I
S  T  R  A  P  R  O  T  E  I  N  S  L  O  A  I  C
C  Y  T  O  K  I  N  E  S  I  S  K  U  L  S  S  R
R  M  I  C  R  O  S  C  O  P  E  S  S  G  M  M  E
V  M  E  K  M  L  I  P  I  D  S  U  I  I  E  N  T
A  D  A  A  D  E  L  B  E  L  I  E  O  B  T  O  I
C  O  R  E  T  N  U  C  L  E  U  S  N  O  A  P  C
U  Y  Z  K  O  R  G  A  N  E  L  L  E  D  P  D  U
O  M  I  C  R  O  T  U  B  U  L  E  S  Y  H  T  L
L  N  O  D  V  R  I  B  O  S  O  M  E  S  A  C  U
E  L  C  H  L  O  R  O  P  H  Y  L  L  K  S  F  M
M  I  C  R  O  F  I  L  A  M  E  N  T  S  E  I  E
```

Word Search

A commonly available type of activity found in bookstores and newspaper stands, words from various disciplines or areas are introduced (with or without hidden messages) and with some degree of artistic license. It's nicer to work on word search activities when it looks inviting! Word search activities help to reinforce the recognition of the words (and hopefully the correct spelling of the words).

The Nervous System

Word Search

The Nervous System—cross out each of words given below. Work vertically, horizontally or diagonally but always in a straight line. The words that remain is a message from the right side of the brain.

```
C S E N S O R Y   M Y E L I N D
  E Y O U A M   H B E T A P A E
C O R T E X O N E U R O N O F N
M   E E   O T D M H A L F N F D
E M F   B N O U I M P U L S E R
N E L I R R R S A A C H   R I
I T E O A E U A P T N E R V E T
N   X N I M   M H E     N   N E
G C   S N E F F E R E N T O T
E P N M Y E   T R A C T   D
S   I S   G   E     H A V E E
  G A N G L I O N B E T T E R
```

Words List

CEREBRUM	MENINGES	IMPULSE
SENSORY	EFFERENT	AXON
MYELIN	AFFERENT	PONS
DENDRITE	GANGLION	NODE
MOTOR	BETA	TRACT
PIA/MATER	IONS	CORTEX
DURA/MATER	HEMISPHERE	NEURON
REM	REFLEX	NERVE
CNS	ACH	BRAIN
EEG		

The Microscope

Word Search

The Microscope—Look for the words given below. Some of the words have been separated (indicated by the slash); work vertically, horizontally or diagonally but always in a straight line.

```
            H  C
            I  O
            G  N
         H  H  D  D
         C  S  E  R
         O  T  N  Y  O  O
         N  A  S  E  E  C  W
         C  G  E  O  T  P  U  O
         A  E  R  S  B  M  I  L
   E  B  V  C  L  E  T  J  T  E  A
   T  U  B  E  L  E  E  R  W  E  T  C  R
      A  R  M  I  N  F  I  N  E  C  S  E
   I  S  B  P  S  L  I  D  E  Y  T  P  O
   M  A     S  N           I  A  I  F  Y
         I  N              G  L  V  G
                          E  L  T  E
                          E  A  H  T
D  I  A  P  H  R  A  G  M     V  N  R  P  O
   B  J                 E  C  E  O  B  I  L
   T                    S  R  S  A  E  O
                        E  S  C  W  M
                        E  E  P  R  E  P
                        M  I  R  R  O  R
L  I  G  H  T  P  A  T  H  A  N  P  O  W  E  R
A  D  J  U  S  T  M  E  N  T  K  N  O  B  S  T
```

Words List

CONDENSER LENS

BASE

TUBE

ARM

STAGE CLIPS

OBJECTIVE

EYEPIECE

PILLAR

DIAPHRAGM

LEVER

LIGHT PATH

SLIDE

STAGE

FINE/ADJUSTMENT KNOBS

NOSE/PIECE

CONCAVE MIRROR

OCULAR

LOW/HIGH/POWER

WET/DRY/PREP

Chemistry

Word Search

Chemistry—Look for the words given below. Work vertically, horizontally or diagonally but always in a straight line.

```
              A   I   N   E   R   T
          N   U   C   L   E   U   S   A
      G   I   O   N   L   I   P   I   D
      A   T   E   X   B   O   N   D   E
      S   C   T   B   Y   L   A   L   S
      H   Y   D   R   O   G   E   N   T
      U   A   T   O   M   E   N   J
      T   S   E   I   O   N
      O   N   E   R   N   L
  L   T                   E   E
  I   S                   R   C
N   G   E   D   N   A      P   T   U   E   T   S
Q   O   E   S   M   M   W   T      O   K   F   N   L   A   I   Y
L U R   P   U   S   A   T   E      T   A   A   E   R   E   E   B
I A B   C   A   T   I   O   N      I   T   R   N   A   C   D   V   P
F R I   A   E   A   R   N   G      S   G   P   R   O   T   O   N   O
E K T   R   D   R   L   O   U      Y   S   A   L   T   R   A   D   K
  E S   O   T   C   I   P   N          C   A   R   B   O   N   A
  O V   Y   H   Y                      A   N   I   O   N   N
```

Words List

ELEMENTS	CARBON	WATER	GAS
MOLECULE	OXYGEN	ORBITS	SOLIDS
ELECTRON	NUCLEUS	BOND	INERT
ATOM	ION	QUARK	DNA
PROTON	ANION	LIPID	RNA
ENERGY	NEUTRON	STARCH	FATS
HYDROGEN	CATION	SALT	LIFE

Energy

Word Search

Energy—Look for the words given below. Some of the words have been separated (indicated by the slash); work vertically, horizontally or diagonally but always in a straight line.

```
                        A   S   E
                    S   P   E   C   I   N
                    E   P   A   N   T   T
                    R   H   R   T   E   I   E
                    N   A   D   O   H   R   V
                            D   W   G   E
                            U   A   Y   E
                            A   C   Y   N   C
                            T   T   Z   O   G
                            P   Y   F   M   T
S           C   Y   C   L   E   M           Y   M   A   E   R
F   A   S   O   E   L   E   C   T   R   O   N   U       E   C   T   A   I
A   D   P   C   Y   T   O   C   H   R   O   M   E   S   F   A   D   S   T   A   N   A   N
D   P   G   L   U   C   O   S   E   R   E   D   U   C   T   I   O   N   O   B   S   S   A
H   A   R   C   A   L   V   I   N   B   E   N   S   O   N   E   N   T   R   O   P   Y   D
R   E   S   P   I   R   A   T   I   O   N   C   A   R   R   I   E   R   S   L   O   S   H   U
G   L   Y   C   O   L   Y   S   I   S   O   X   I   D   A   T   I   O   N   I   R   T   O   R
O   K   R   E   B   S   T   H   E   R   M   O   D   Y   N   A   M   I   C   S   T   E   F
L               U   S   I               T   S   G   O   O   M   D   M
```

44

Words List

ENERGY

METABOLISM

THERMODYNAMICS

ENTROPY

ENZYMES

COFACTORS

CARRIERS

PATHWAY

PRODUCT

ACTIVE/SITE

PH

ATP/ADP

CYTOCHROMES

ELECTRON/TRANSPORT/SYSTEM

OXIDATION

REDUCTION

NADH

FADH

FAD

NAD

CALVIN BENSON/CYCLE

RESPIRATION

KREBS/CYCLE

GLYCOLYSIS

GLUCOSE

The Cell

Word Search

The Cell—Look for the words given below. Some of the words have been separated (indicated by the slash); work vertically, horizontally or diagonally but always in a straight line.

```
                        I   C   B
                T   S   C   E   R   D   N   A           B   N   M   E   O
                T   E   A   I   L   O   S   Y   T   V   I   A   F   I   U   G   R
            A   U   C   E   L   L   W   A   L   L   E   N   C   R   C   K   N   E   O
            M   U   T   W   I   T   N   H   I   C   S   U   T   M   R   A   L   H   E
        P   Y   E   L   E   A   H   B   U   L   K   I   C   E   A   O   R   E   C   N
        R   L   N   U   C   L   E   O   L   U   S   C   L   R   T   S   Y   E   H   D   C
        O   O   D   E   H   T   O   P   O   R   E   L   E   I   R   C   O   U   R   O   H
        T   P   O   X   R   O   R   F   L   O   W   E   U   A   I   O   T   W   O   C   R
    N   E   L   P   O   O   C   Y   T   O   P   L   A   S   M   X   P   E   E   M   Y   O
    U   I   A   L   C   M   E   M   I   C   R   O   B   O   D   I   E   S   N   O   T   M
    C   N   S   A   Y   A   N   U   S   R   O   U   G   H   E   R   E   L   H   S   I   O
C   L   M   T   S   T   T   A   N   D   T   L   I   P   I   D   W   Y   O   O   C   P
J   H   E   I   S   M   I   I   R   N   A   N   O   M   E   T   E   R   H   S   E   M   M   L
U   L   A   T   S   I   C   N   I   H   I   C   P   O   L   A   R   H   L   O   K   E   T   A
N   O   R   O   M   C   E   E   O   O   C   H   L   O   R   O   P   L   A   S   T   S   O   S
C   R   E   C   O   R   N   N   L   O   E   T   A   T   E   R   A   T   P   O   R   S   C   T
T   O   N   H   O   E   E   Z   E   K   N   O   S   T   M   I   L   L   I   M   E   T   E   R
I   P   V   O   T   T   R   Y   C   E   L   L   M   E   M   B   R   A   N   E   T   B   O   T
O   H   E   N   H   I   G   M   O   G   O   L   G   I   B   O   D   Y   U   S   A   R   U
N   Y   L   D   E   C   Y   E   V   A   C   U   O   L   E   S   B   E   B   S   G   B
    L   O   R   R   U   E   S   B   I   L   A   Y   E   R   O   O   T   A   A   U
    L   P   I   P   L   A   S   M   I   D   S   C   A   U   M   D   L   N   L   S
    E   O   E   U   P   R   O   K   A   R   Y   O   T   E   Y   E   E   I
        N   A   M   I   C   R   O   M   E   T   E   R   S   L   S   M   N
            O   T   N   U   C   L   E   O   P   L   A   S   M
                A   C   T   I   V   E   S   I   T   E
                    O   S   M   O   S   I   S
```

46

Words List

CELL THEORY

CYTOPLASM

PROTOPLASM

CELL MEMBRANE

NUCLEUS

CENTRIOLE

CHLOROPLASTS

CELL WALL

NUCLEOLUS

LYSOSOMES

MICROBODIES

RIBOSOMES

PROTEIN

PLASMIDS

GOLGI BODY

MICROSCOPE

MICROMETER

NANOMETER

MILLIMETER

HOOKE

BROWN

LEEUWENHOEK

PROKARYOTE

EUKARYOTE

BACTERIA

ORGANELLES

MITOCHONDRION

VACUOLE

ROUGH ER

SMOOTH ER

CHROMOSOMES

NUCLEOPLASM

PORE

CHROMATIN

EXOCYTIC/VESICLE

ENDOCYTIC/VESICLE

NUCLEAR ENVELOPE

ENZYMES

AMYLOPLASTS

CHLOROPHYLL

CHROMOPLAST

CILIA

TUBULES

BASAL/BODY

MATRIX

JUNCTION

ENDOPLASMIC RETICULUM

LIPID/BILAYER

POLAR

BULK/FLOW

OSMOSIS

ENERGY

ACTIVE SITE

ATP

DNA

RNA

MTOC (Microtubule organizing center)

The Plant and Animal Kingdoms

Word Search

The Plant and Animal Kingdoms—Look for the words given below. Work vertically, horizontally or diagonally but always in a straight line.

```
E                       O  S                       E
C  V                    A  S                    K  F
H  M                    M  O                    L  H
I  A  C              E  O  R  R                 E  W  S
N  M  Y  R              M  E  U  F           A  O  L  P
O  M  C  O  L        Y  O  B  S  U  L        R  I  T  O
D  A  A  T  Y  I     B  N  A  V  N  I     M  M  B  O  N
E  L  D  I  C  O  C  E  F  E  I  D  G  O  P  S  E  T  R  A  G
R  S  P  F  O  E  A  E  B  R  N  I  I  N  R  M  W  E  Y  D  E
M  F  R  E  P  R  I  N  U  A  D  C  P  R  O  T  O  Z  O  A  N
S  V  I  R  O  I  D  S  F  N  C  O  A  L  T  F  C  L  P  N  A
H  O  O  S  D  U  E  O  R  S  A  T  D  T  I  R  H  I  H  G  R
Y  B  N  S  S  X  E  E  L  P  S  E  V  S  O  O  C  Y  I  T
P  I  G  F  R  O  N  D  U  E  E  S  F  R  T  G  R  H  T  O  H
H  V  A  R  C  H  A  E  B  A  C  T  E  R  I  A  D  E  E  S  R
A  A  M  P  H  I  B  I  A  N  S  W  R  O  A  A  A  N  S  P  O
E  L  D  I  A  T  O  M  C  F  L  Y  N  N  N  D  T  E  P  E  P
E  V  O  A  E  C  H  Y  T  R  I  D  S  M  S  I  E  M  I  R  O
U  E  G  N  A  L  G  A  E  N  G  F  H  O  A  D  S  A  D  M  D
G  Y  M  N  O  S  P  E  R  M  S  I  O  L  P  N  W  T  E  S  S
L  J  F  E  W  S  Q  U  I  D  E  S  F  L  A  T  W  O  R  M  E
E  E  I  L  E  E  C  H  A  C  A  H  A  U  M  E  T  D  O  O  A
N  L  S  I  N  S  E  C  T  S  B  N  E  S  C  O  N  E  S  T  G
I  L  H  D  W  H  A  C  H  I  T  O  N  K  T  A  W  S  A  H  L
D  Y  Y  S  T  B  I  R  D  S  A  N  T  S  R  E  P  T  I  L  E
```

Words List

FOX	WORMS	MAMMALS	BRYOPHYTES
TOAD	MAN	CHORDATES	ALGAE
ANTS	APE	ECHINODERMS	CHYTRIDS
LION	CONES	CHITON	FUNGI
BEAR	DICOTS	LEECH	AMOEBA
EAGLE	CYCAD	INSECTS	PROTISTANS
DOG	FROND	ARTHROPODS	PROTOZOAN
CAT	SORUS	ANNELIDS	SLIME MOLDS
FROG	LICHEN	MOLLUSKS	EUBACTERIA
LICE	HYPHAE	NEMATODES	ARCHAEBACTERIA
MOTH	DIATOM	FLATWORM	BACTERIA
FLEA	ROTIFERS	SPONGE	PRION
FLY	JELLY/FISH	SEED PLANTS	VIROIDS
SPIDER	EUGLENID	GYMNOSPERMS	VIRUSES
SQUID	BIVALVE	ANGIOSPERMS	MONERANS
FISH	AMPHIBIANS	FERNS	
BIRDS	REPTILE	LYCOPODS	

The Plant World

Word Search

The Plant World—Look for the words given below. Work vertically, horizontally or diagonally but always in a straight line.

```
            A  C  Y  P  R  E  S  S  O  M  E
         S  P  U  F  L  I  V  E  R  W  O  R  T
      A  N  P  U  F  F  B  A  L  L  S  S  C  A  B
   C  O  T  N  A  L  I  L  Y  B  D  E  S  P  I  N  E
B  O  L  G  C  W  N  G  K  A  M  O  L  D  S  E  H  R
P  I  N  I  M  A  E  A  I  N  U  R  D  A  F  Q  I  A  N
E  R  I  C  A  C  E  M  Y  S  C  S  D  S  K  U  Y  O  F
A  C  F  H  P  T  D  A  H  H  H  G  E  G  R  A  S  S  E  S
C  H  E  E  L  U  N  R  I  M  O  M  R  T  M  W  E  I  R
H  N  R  N  E  S  O  D  M  T  A  C  O  R  N  R  W  K  N
E  L  B  R  O  O  M  R  A  P  E  Y  A  S  L  O  N  E  S
L  E  G  U  M  E  I  C  G  F  A  B  A  S  S  O  D  L  N
   T  H  S  B  E  A  N  N  E  I  A  E  P  H  T  A  P  L
      S  Q  U  A  S  H  O  P  P  R  O  E  P  I  T  N
         M  P  N  G  A  L  F  A  L  F  A  C  L  E
            U  O  F  M  I  N  T  E  W  H  A  H  E
               T  P  L  A  T  C  Y  T  O  M
                  P  O  T  A  T  O  E
                     Y  W  S  N  M
                        S  E  A  A
                        O  K  R  T
                        Y  U  D  O
                        B  D  O  T
                        E  Z  G  H
                        A  U  W  O
                        N  V  O  L
                     V  I  O  L  E  T
                  U  R  N  D  Y  A  L  T
            O  S  T  O  N  E  C  R  O  P  M  E
```

50

Words List

SPANISH MOSS	LEGUME	ALFALFA	PEACH
CORN	CONIFER	SQUASH	STONECROP
PUFFBALLS	CONES	POTATO	DATE
MUSHROOMS	LIVERWORT	TOMATO	WEED
MOLD	FERNS	BARLEY	ASH
SQUAWROOT	MOSS	PEA	BEAN
DODDER	MAPLE	CYPRESS	HOLLY
BROOMRAPE	CACTUS	SOYBEAN	KELP
FUNGI	LILY	BEECH	SCAB
ORCHID	BANYAN	KUDZU VINE	SMUT
ALGAE	MAGNOLIA	VIOLET	FIR
LICHEN	PINE	DOGWOOD	POPPY
OAK	BIRCH	MINT	FIG
GRASSES	SUNFLOWER	MAYAPPLE	

Ecology

Word Search

Ecology—Look for the words given below. Work vertically, horizontally or diagonally but always in a straight line.

```
                              H
                              B  G
                           I  E  C  R
                        O  O  L  O  N  O
                     S  P  E  C  I  E  S  W
                  O  P  E  G  Y  C  L  I  M  A  T  E
               C  H  R  A  D  H  C  L  I  M  A  X  H  F
            E  E  A  E  S  E  S  U  C  E  S  S  I  O  N  A
         A  R  M  B  D  O  C  C  O  N  S  U  M  E  R  S  F  L
      N  E  E  A  I  A  N  O  P  O  P  U  L  A  T  I  O  N  N  L
            C  T  T  P  M  R  R  L  G  Y  R  E  O  A
            A  A  I  A  P  O  S  E  O  T  H  D  E  S
            R  T  O  R  O  D  H  D  Y  G  W  O  O  B
            B  D  N  A  S  U  D  U  E  E  Y  I  S  I
            O  I  E  S  E  C  T  T  B  N  L  H  E  O
            N  S  O  I  R  E  C  O  S  Y  S  T  E  M
            S  E  R  T  S  R  W  A  T  E  R  I  T  E
            P  A  A  I  S  C  Y  C  L  E  U  T  N
            R  S  I  S  D  C  O  M  M  U  N  I  T  Y
            E  E  N  M  S  H  A  D  O  W  Y  A  O  F
            D  E  T  R  I  T  I  V  O  R  S  B  Z  T
            A  O  R  C  O  M  P  E  T  I  T  I  O  N
            T  M  I  M  I  C  R  Y  G  A  N  O  N  I
            O  C  A  M  O  U  F  L  A  G  E  T  E  S
            R  L  A  K  E  M  U  T  U  A  L  I  S  M
            H  E  R  B  I  V  O  R  S  M  I  C  N  T
```

52

Words List

ECOLOGY	COMMUNITY	PREDATOR	DDT
POPULATION	SPECIES	CAMOUFLAGE	SEASON
PRODUCERS	DENSITY	MIMICRY	GYRE
CONSUMERS	GROWTH	SUCESSION	RAIN/SHADOW
DECOMPOSERS	PREDATION	CLIMAX	BIOME
DETRITIVORES	PARASITISM	CLIMATE	SOIL
ECOSYSTEM	DISEASE	FOODWEB	ZONES
BIOSPHERE	COMPETITION	HERBIVORS	FALL
HABITAT	ABIOTIC	WATER/CYCLE	OCEAN
NICHE	MUTUALISM	CARBON	LAKE
BIOTIC	PREY		

Evolution

Word Search

Evolution—Look for the words given below. Work vertically, horizontally or diagonally but always in a straight line.

```
E V R               U S L D           T P O       T O Y
V I E               T O T I           C H O       H E A
O R C L             B U U V           V E E       Y R M
U U O L K           A P P E           R N B       L T
  S R T I U         L B R R           E O C       L H
  I D O N F         A O A G       A F T       C A N
    I N G I E       N T D E       U O Y   N R M D
    S S D S R       C T I N       S U P T D I S
      U O C A A     E L A C     E H N E Y M P U
        M L D D     D E T E     Y E D W I E P
          T A A F F S N I     B C R E C C P
            S P T I   E E O   R H O I R I Y L
              S T A I   I L C N   I W S N Y E T A
                I X C O C E K   D N S B V S F F O
                V A R A N C S F D L E L E T F O
                E V O L U T I O N R D E R E E A
                G P A A R I G S G M I U G P C
                  O L R O O E S N U F F E D T
                  L L P I N N I R T F I T
                  Y E H A A O L E A E C
                  M L Y N T T L P T R
                  O E L I U Y I R I E
                  R V A N R P N O O N
                  P I G F A E E D N T
                  H A E E L T R U E I
                  I B N R N O A C L A
                  S I E T L E W T O L
                  M L T I Y Z D I R R
                  U I I L N O E O D S
                  T T C I A N N N E D
                  W Y H T G E N E R A
```

Words List

EVOLUTION

VARIATION

HARDY-WEINBERG

MUTATION

NATURAL/SELECTION

GENE/FLOW

GENETIC/DRIFT

FOUNDER EFFECT

ALLELE

GENERA

CLASS

ORDER

FOSSIL/RECORD

LINE

SOUP

ERA

LIFE

VIRUS

SPECIES

HYBRID

INFERTILITY

VIABILITY

GENOTYPE

PHENOTYPE

TAXA

PHYLA

KINGDOM

BOTTLENECKS

ADAPTIVE/RADIATION/ZONE

DIFFERENTIAL/REPRODUCTION

MIMICRY

BALANCED/POLYMORPHISM

DIVERGENCE

CONVERGE

ISOLATION

Protein Synthesis

Word Search

Protein Synthesis—Look for the words given below. Work vertically, horizontally or diagonally but always in a straight line.

```
                              P  G  D
                           C  R  E  T  N
                        I  Y  O  N  R  G
                        A  T  T  E  A  L  A
                        I  M  O  E  T  N  Y  A                    B  B  I
            T           S  A  I  S  I  I  S  C  N  S           P  A  O  C
   A  C  T  I  V  A  T  O  R  N  I  N  C  L  I  T  O  E     T  H  I  R  D
   O  F  H  M  U  T  A  T  I  O  N  S  S  A  N  I  N  T  R  O  N  A  R  Y
   N  P  R  O  M  O  T  E  R  A  E  Y  G  T  E  C  S  H  O  I  D  U  C
      L  D  O  P  E  R  O  N  C  O  N  R  I  B  O  S  O  M  E  N  L
         T  R  A  N  S  C  R  I  P  T  I  O  N  D  C  C  N  U  O  E
         R  R  W  H  E  O  N  D  L  H  T  N  H  O  I  I  N  N  G
            R  A  S  A  D  R  S  A  E  E  G  O  N  N  I  E  N  G
            T  E  N  W  O  M  E  S  S  E  N  G  E  R  R  N  A
            E  G  P  S  N  U  Y  M  I  T  H  E  T  T  U  G  O
               N  U  R  F  S  R  A  S  H  O  U  H  C  E  P
               S  Z  A  E  E  U  A  L  D  M  Y  L  N  I  I
               P  Y  N  S  R  T  C  O  M  E  E  R  H
                  U  M  I  S  R  R  I  U  I  A  S
                  F  E  N  O  N  S  L  T  R
                  E  F  P  E  R  A  O  E
                  S  A  T  R  I
```

Words List

PROTEIN SYNTHESIS	ANTICODON	GENETICS	ENZYME
TRANSCRIPTION	ADENINE	SERINE	BARR/BODY
TRANSLATION	THYMINE	GLYCINE	PUFFS
RIBOSOME	GUANINE	MUTATION	TUMOR
MESSENGER RNA	URACIL	REPRESSOR	PLASMA
NUCLEUS	CYTOSINE	OPERATOR	CLONE
AMINO ACIDS	INTRON	ACTIVATOR	GENE
TRANSFER RNA	PROMOTER	OPERON	CODON

The Circulatory System

Word Search

The Circulatory System—Look for the words given below. Some of the words have been separated (indicated by the slash); work vertically, horizontally or diagonally but always in a straight line.

```
                                P   H     A   C   T
      B   I   T                 A   E   O   R
      O   V   A   L   E         C   A   R   E
      A   O   R   T   A   S     E   R   V
              N   R   E   L     M   T
              R   U   E         A   E
              A   L   U   M   D   I   K   D   S   E   R
          L   H   U   M   E   U   T   E   E   T   R   O   M
      F   I   B   R   I   N   C   I   R   M   R   T   S   O   C
      C   P   I   E   T   A   T   R   I   A   O   H   U   N   Y
      E   I   L   A   N   S   U   O   E   H   K   E   P   O   C
      L   D   O   E   S   I   N   D   D   E   P   E   C   L
      L   A   M   O   T   C   A   P   I   L   L   A   R   Y   E
      S   I   A   B   H   T   R   I   A   T   O   R   R   T
      A   R   T   E   R   Y   T   L   S   E   B   I   I   E
          C   A   S   O   V   E   U   T   I   K   N   O   H   K
          N   I   M   E   R   B   O   S   N   G   R   E   V
          G   T   B   N   I   B   L   O   O   D   V   M   E   V
          I   Y   U   U   O   V   E   I   N   S   E   O   N   O
          N   A   S   L   S   I   N   U   S   A   N   G   T   L
          A   D   E   U   Z   I   N   C   N   A   L   R   U
              I   S   Y   S   T   O   L   E   T   C   O   I   M
                  V   A   L   V   E   Y   M   I   A   B   C   E
                  S   H   O   C   K   M   B   G   V   I   L   L
                      D   U   B   B   P   O   E   A   N   E
                      C   L   O   T   H   L   N   I   E   S
                          B   I   T   U   O   N   T
                              I   S   E
```

58

Words List

VOLUME	EKG	FIBRIN	PACEMAKER
OVALE	STROKE	RED/BLOOD/CELLS	DUCTUS ARTERIOSUS
SINUS	SHOCK	HEMOGLOBIN	AORTA
LUMEN	LIPID	LYMPH	ATRIA
UREA	OBESITY	ARTERY	ANEMIA
ZINC	EMBOLUS	EDEMA	CAPILLARY
IRON	ANGINA	VENULE	HEART
CLOT	VALVE	THROMBUS	SUPERIOR VENA CAVA
LDL	LUBB	SERUM	SYSTOLE
HDL	DUBB	HEPARIN	DIASTOLE
MONOCYTE	CYCLE	ANTIGEN	VEINS
VENTRICLES			

The Digestive System

Word Search

The Digestive System—Look for the words given below. Some of the words have been separated (indicated by the slash); work vertically, horizontally or diagonally but always in a straight line.

```
        W  E  A
        H  N  N              A  C  I  D
        E  U  Z           N  E  N  T  A  C  D  K
        S  O  Y        M  E  T  A  B  O  L  I  S  M
        F  P  M  R  O  O  B  L  D  I  G  N  P
        A  P  E  T  O  O  L  I  V  E  R  A
        N  E  C  S  N  L  F  N  A  S  A  R  T
        C  P  T  T  H  I  A  T  B  T  B  A  H
        R  S  U  O  E  S  T  E  S  I  D  S  C
        E  I  M  M  M  S  S  O  O  O  T  L
        A  N  B  A  O  D  A  T  R  N  M  A  B
        S  U  L  C  Y  L  E  I  P  A  E  L  M
        I  L  T  H  I  E  C  N  T  P  N  S  R
     I  S  C  A  V  T  I  A  E  I  P  I  I  M
     P  O  R  E  A  H  O  T  L  S  O  C  A  S  T
  T  A  C  N  T  R  T  O  N  R  O  P  N  N  C  R
E  S  A  L  T  S  P  H  A  R  Y  N  X  M  G  A  R  R  C  D  I  E
M  I  N  E  R  A  L  Y  M  O  G  M  B  E  R  U  C  I  O  H  I  N
C  O  V  I  T  A  M  I  N  L  M  O  U  T  H  E  T  E  T  E  X  H
H  B  A  T  H  E  A  T  A  I  T  T  F  O  L  I  C  S  E  W  S
O  E  M     O  U  L  N  D  B  M  I  C  E  L  L  E  D  I  O
L  S  I        N  D  E  V  I  L  L  U  S  B  I  L  E  N
I  I  N        S  E  C  R  E  T  I  O  N  C  A  R  N
N  T  O        E  B  I  O  T  I  N  F  C
E  Y  L  Y     U  L  Y  N  S
```

60

Words List

METABOLISM	PANCREAS	VILLUS	MINERAL
DIGESTION	ANUS	CHYME	OBESITY
ENZYMES	SECRETION	ULCER	BIOTIN
ANABOLISM	ABSORPTION	ABDOMEN	FOLIC/ACID
STOMACH	MOTILITY	HCl	AMINO/ACID
LIVER	PARASTALSIS	MICELLE	BILE/SALTS
SMALL/INTESTINES	TEETH	APPENDIX	CHEW
COLON	TONGUE	CALORIES	TRACT
RECTUM	MOUTH	FATS	CHOLINE
SALIVARY GLANDS	PEPSIN	BMR	NIACIN
PHARYNX	PROTEIN	COVITAMIN	

The Endocrine System

Word Search

The Endocrine System—Look for the words given below. Work vertically, horizontally or diagonally but always in a straight line.

```
            E  P  I  N  E  P  H  R  I  N  E  M  E  S  S  E  N  G  E  R
               H  Y  P  O  A  C  T  I  V  E
               P  I  T  U  I  T  A  R  Y
               O  N  C  E
               T  H  U
               B  H  I  T
            C  A  A  B  E
            T  R  L  I
      T  S  M  A  L  P  O  A  A  T  H  Y  R  O  I  D  H  T
   B  G  T  S  M  A  A  E  D  M  I  N  S  U  L  I  N  T  E  S
T  O  L  E  H  B  M  L  R  R  U  N  I  U  C  O  R  T  I  S  O  L  A
C  K  U  R  L  P  I  E  F  E  S  G  L  G  O  R  C  B  L  T  V  I  T  P
O  I  C  O  U  N  N  O  S  N  E  F  E  A  L  O  E  A  L  E  A  S  H  I  P
R  O  O  I  S  A  E  S  H  A  A  A  S  R  D  T  L  T  H  S  R  L  Y  N  T
T  D  S  D  L  T  F  O  L  L  I  C  L  E  A  T  L  E  I  G  Y  E  M  E  H
E  I  E  S  F  I  R  S  T  I  S  T  I  M  U  L  A  T  I  N  G  T  U  A  G
X  N  S  P  R  O  T  E  I  N  H  O  R  M  O  N  E  R  O  H  N  S  S  L
   E  K  K  S  G  O  I  T  E  R  R  H  Y  P  E  R  A  C  T  I  V  E
      M  E  D  U  L  L  A  P  A  N  C  R  E  A  S  O  F  T  S
         C  N  T  H  Y  R  O  X  I  N  E  G  G  S  E
         S  O  M  I  N  A  C  U  T  I  V  E  B
```

62

Words List

THYROXINE

IODINE

THYMUS

PTH

MSH

LTH

GH

FSH

MEDULLA

CORTEX

AMINE

PROTEIN

AMP

FIRST/MESSENGER

ISLETS

TESTES

BLOOD/SUGAR

PANCREAS

GLUCOSE

CRETIN

INSULIN

BETA/CELL

CORTISOL

HYPERACTIVE

HYPOACTIVE

OVARY

ADRENALS

STEROIDS

PINEAL

EPINEPHRINE

PITUITARY

HYPOTHALAMUS

INHIBITING FACTOR

FOLLICLE/STIMULATING/HORMONE

THYROID

ADRENALINE

GOITER

The Skeletal System

Word Search

The Skeletal System—Look for the words given below. Some of the words have been separated (indicated by the slash); work vertically, horizontally or diagonally but always in a straight line.

```
                    P  N  T  A  D  B  A  S  E
                 V  P  A  E  P  I  N  C  U  S  O  R
              L  C  O  E  S  N  P  S  D  F  M  B  R  A  H
           T  I  H  M  L  A  D  E  L  S  U  O  T  B  D  E  M
        F  U  G  I  E  V  L  O  N  O  K  N  L  O  I  I  A  E
     S  L  M  A  T  R  I  X  N  D  C  E  E  T  A  T  U  D  T  A  K
     T  E  A  M  I  C  S  A  X  I  A  L  D  I  V  I  S  I  O  N  F
     I  X  R  E  N  L  I  M  B  C  T  E  G  O  U  T  V  A  S  S  O
     B  I  R  N  E        U  I  T           U  P  T  H  S
     I  O  O  T           L  O  O                 H  E  A  S
     A  N  W  S           E  A  N  N              Y  O  F  A
        S  A  Y  H  I  P  V  H  R     Q  U  F  B  U  R  S  C  T
           N  N  C  N  E  T  A  I     U  E  R  U  L  V  I  Y
           V  O  H  R  H  O  N        L  A  R  C  L  S  T
           I  V  S  L  A  E  D        B  C  S  I  U  C  E
           L  I  P  I  V  O  T        O  T  A  T  L  K
              A  R  P  E  V  U  D  O  N  W  U  U  D  N
              L  A  E  R  J  O  I  N  T  S  R  I  E  A
              I  F  S  N  B  V  I  C  E  E  S
              N  E  I  E  P  I  P  H  Y  S  I  S
              Y  M  A  C  T  S  A  C  R  U  M  P
              O  U  N  K  H  I  N  G  E  M  S  I
              A  R  T  H  R  O  L  O  G  Y  I  N
              C  V  O  L  A  N  T  H  I  G  H  E
                 S  L  I  D  E  K
```

64

Words List

TOE	ULNA	LIGAMENTS	APPENDICULAR/DIVISION
FOSSA	TIBIA	MATRIX	AXIAL DIVISION
BASE	HAVERSIAN	SPRAIN	SKELETON
LIMB	MOLT	FEMUR	OSTEOCYTE
SPINE	CHITIN	PIVOT	DIAPHYSIS
HIP	VOLANT	PELVIS	SUTURE
SHAFT	SLIDE	BONE	EPIPHYSIS
HEAD	TENDON	MARROW	ARTHROLOGY
NECK	GOUT	SYNOVIAL	JOINTS
THIGH	NASAL	BURSA	CILIA
HAND	ORBIT	FRACTURES	HINGE
FOOT	VOMER	FLEXION	DISLOCATION
KNEE	ANVIL	SACRUM	LEVERS
ELBOW	INCUS	RADIUS	

The Reproductive System

Word Search

The Reproductive System—Look for the words given below. Some of the words have been separated (indicated by the slash); work vertically, horizontally or diagonally but always in a straight line.

```
                  G O N A D H S I M O R U L A
                  P H E N O T Y P E M O N O P E
              R A I O S E M E N B V F S A O I E C K
          Q E Z L I T R I C F E R T I L I Z A T I O N
        L U C Y P T O C N S G M A I L L S Z F L A B I A
      T U M E G E N E S A G P M Y L D P A A L E E L E S
  T E S T E S O N A S Y N N E A E T E I K M G A M E T E
  H Y M E N S T I G M A T N O N A R F L O W E R L A L U
  H O W R S I E S A S N E I T S M C H O R I O N C M L L
  A B U E V U L K C T M U P C R O S S N I N S T Y L E
    O S S E E C O R H B H U O I D U C T V A G I N A
        M         O E R E N T M     T B I R T H E
                  T R Y T N G P         G R E
                  U E O E E S L
                  M V T R T E A
                M U P M O T R N
                L T I C Z S T T T
              E A N S O Y Q O A T
              H A E T W G U L T L
              L A S I P O A I I T
                R U L E U R C O
                D O U R S E E N
                A G S L N L
                  M L D U L
                  N A S
                  O I N
                  M E O D
                  T N
```

Words List

FERTILIZATION	PUNNETT SQUARE	UTERUS	AMNION
HYBRID	ZYGOTE	PENIS	MORULA
POLLEN TUBE	HETEROZYGOUS	COWPER'S/GLAND	BIRTH
PETAL	GENES	VAGINA	IUD
STAMEN	DOMINANT	CROSS	MENSES
FILAMENT	RECESSIVE	SEMEN	HYMEN
ANTHER	GAMETE	SERTOLI CELL	LABIA
FLOWER	SPERM	PISTIL	FEMALE
STYLE	EGG	IMPLANTATION	MALE
STIGMA	GERMINAL/CELL	CHORION	DUCT
OVULE	TESTES	EMBRYO	GONAD
PHENOTYPE	SCROTUM		

Heredity

Word Search

Heredity—Look for the words given below. The words, however, do not appear horizontally, vertically or diagonally in a straight line. In most cases the words will break abruptly. For example, the word MUTATIONS has already been identified and circled for you.

```
I     Y  S  R  T  E  N  E  G  Y        R  F  L
N     N  M  E  I  C  C  O  U  D     R  D  N  O  P
H  M  D  U  N  I  L  E  S  N  I  A  O  A  C  H  A
E  U  R  T  R  U  T  N  G  I  O  N  P  L  A  S  I
R  T  O  R  A  D  I  A  T     L  E  D  O  W  N  S
I  A  M  E  T        S  P  U  T  R        Y
T  G  L  B  I  N  O     O  S        I  R  X  N
A  A  E  O        R     O  V  E  R  S  A  P  D
N  X  M  N  N  U  C  L  E  O     Y  M  O  Y  H  R
C  L  N  E  S  O     I     T  I  D  E  S  E  O
E  I  I  T  S  N     N     R  W  I  U  L  G  N  M
   N  O  N  I  D     K  A  G  E  L  P  E  A  L  E
   K  C  E  S  I     N  S  N  O  S  L  T  L  Y  A
G  E  N  E  J  S  S  D  I  S  O  C  I  I  A  K  I
   D        U  L  R  E  D  R  A        O  C  E  R
   C  P     N     O  C  A  T  I  O  N  N  T  T  U
   L  O     C        I  A  I  M  E  S  O  O  N
   E  L     T        O        H  K  L  E  I  N
   F  Y     I     N           E  R  E  T  L  E  F
   T  P  L  O  I  D  Y     M  S  Y  N  D  R  O  M
   L  I  P  N           O  P  H  I  L  I  A     E
```

Words List

INHERITANCE
MUTATIONS
NUCLEOTIDES
RADIATION
MUTAGENS
ALBINO
X-RAYS
LINKAGE
DELETION
DUPLICATION
TRANSLOCATION
NONDISJUNCTION
POLYPLOIDY
ANEUPLOIDY
CROSSOVER
X-LINKED GENE
GALACTOSEMIA
ACHONDROPLASIA
HEMOPHILIA
DOWN SYNDROME
TURNER SYNDROME
PHENYLKETONURIA
KLINEFELTER SYNDROME
CLEFT LIP
WILSON'S DISEASE
AMNIOCENTESIS
TRISOMY
RFLP
GENETIC COUNSELING

PART FOUR

Sentence Search

A variation on the word search activity sentence search requires that you attempt to find a pattern among letters that will spell out specific words that related to one another thus disclosing a complete message when solved. A series of interesting and, in some instances, little known facts about human biology, world population, botany, birth rate, etc. Just keep looking! Remember, answers are provided for you in the back of the book.

A Hair Raising Tale

Sentence Search

A Hair Raising Tale—Moving from letter to letter up, down, sideways or diagonally here's an interesting piece of information for the gentlemen. Ladies, be aware of this information too. Each letter must be used only once.

```
                                                Y
                                            T
                                        R
                                    O
                                F
                            G   E
                        T   A   T
                    A   E   H
                V   E   M   O
                A   H   E   R
            E   L   L   A   H
        H   R   I   A   I
        S   S   H   A   R
    E   L   E   H   A   M
I   R   T   Y   T   N
H   A   T   A   T   A   H
T   G   H   E   S   O   S
E   C   S   I   H   N
```

Why Is the Biology Department On the Third Floor?

Sentence Search

Why is the Biology Department on the Third Floor? Moving from letter to letter up, down or sideways you'll be glad to discover the answer to our question.

```
                S   A   T   J
                E   L   I   O
                I   C   M   H
        T   U   D   U   B   N
        S   R   Y   O   A   S
        O   I   Y   O   D   H
    N   G   T   A   O   U   D   O
    I   H   S   T   T   R   S   P
    D   C   N   D   S   L   F   K
C   O   R   A   O   E   F   I   O   I
C   T   Y   E   C   E   S   R   U   N
A   I   S   R   E   V   I   N   U   S
```

What Did You Say Again?

Sentence Search

What Did You Say Again? Moving from letter to letter up, down, sideways or diagonally you'll hear what I said.

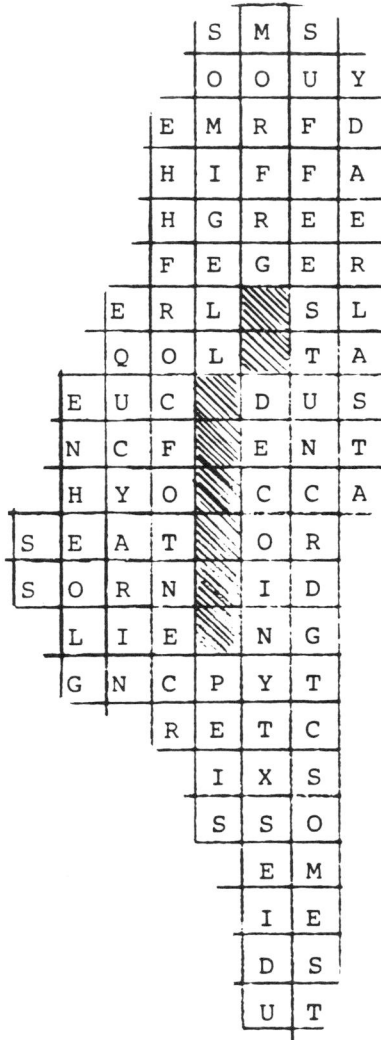

S	M	S		
O	O	U	Y	
E	M	R	F	D
H	I	F	F	A
H	G	R	E	E
F	E	G	E	R
E	R	L	▨	S	L
Q	O	L	▨	T	A
E	U	C	▨	D	U	S
N	C	F	▨	E	N	T
H	Y	O	▨	C	C	A
S	E	A	T	▨	O	R
S	O	R	N	▨	I	D
L	I	E	▨	N	G
G	N	C	P	Y	T
R	E	T	C
I	X	S
S	S	O
E	M
I	E
D	S
U	T

73

Who's Got B and O?

Sentence Search

Who's got B and O?—No, it's the letter B and an outline of the eyeball that you're looking at. Moving from letter to letter up, down, sideways or diagonally each of the symbols would most certainly "B" an "EYE" opener for you.

```
        O   M   L   I   V
        E   E   N   O   L   E
        P       N   G   E
        Y       N   A   R
        T   B   L   B   H   T
        D   O   O   L
        N   E   M   O   O   D
        R   W   O   E   H   T
        O           O   T   Y
        F           P   N   P
        E   I   E   P   E   E
        U   S   T   O   M   B
        R   T   I   S
```

```
            F   O   S   E   L   C
        Y   Y   O   U   H   E   M   U   S   Y
    S   E   E   R   T           S   A   D   A
    M   O   V   E               E   M   I   T
    U   H   E   N   O   H   O   A   N   D
        N   D   R   E   D   T   U   S
```

74

A New Twist

Sentence Search

A New Twist—Oh humm, you've been searching for sentences by moving up, down and sideways and it's time for a change. This time, move from letter to letter only on a diagonal. The exclamation point is the clue.

```
                        R
                  M          E
            F        N          T
         O       I       G        L
      R      T       A      I        A
   Y     A       P      N      S        C
A     L     C       D      A       I
   W     U       I       M      D
      N          !       E
         I     G
            N
```

This Letter Makes Sense

Sentence Search

This Letter Makes Sense—Move from letter to letter up, down or sideways and you'll learn from the title.

```
R  L  I  L  I  E  S
E  E  H  T  E  R  A
T  F  I
A  S  R
W  T  T  R  U
D  M  L  F  E
N  A  O
A  G  W
S  N  E
A  O  R
I  L  S
```

Life Is Dense

Sentence Search

Life is Dense—Moving from one square to another length, width and depth you'll appreciate the density of the information. Every letter will be used once.

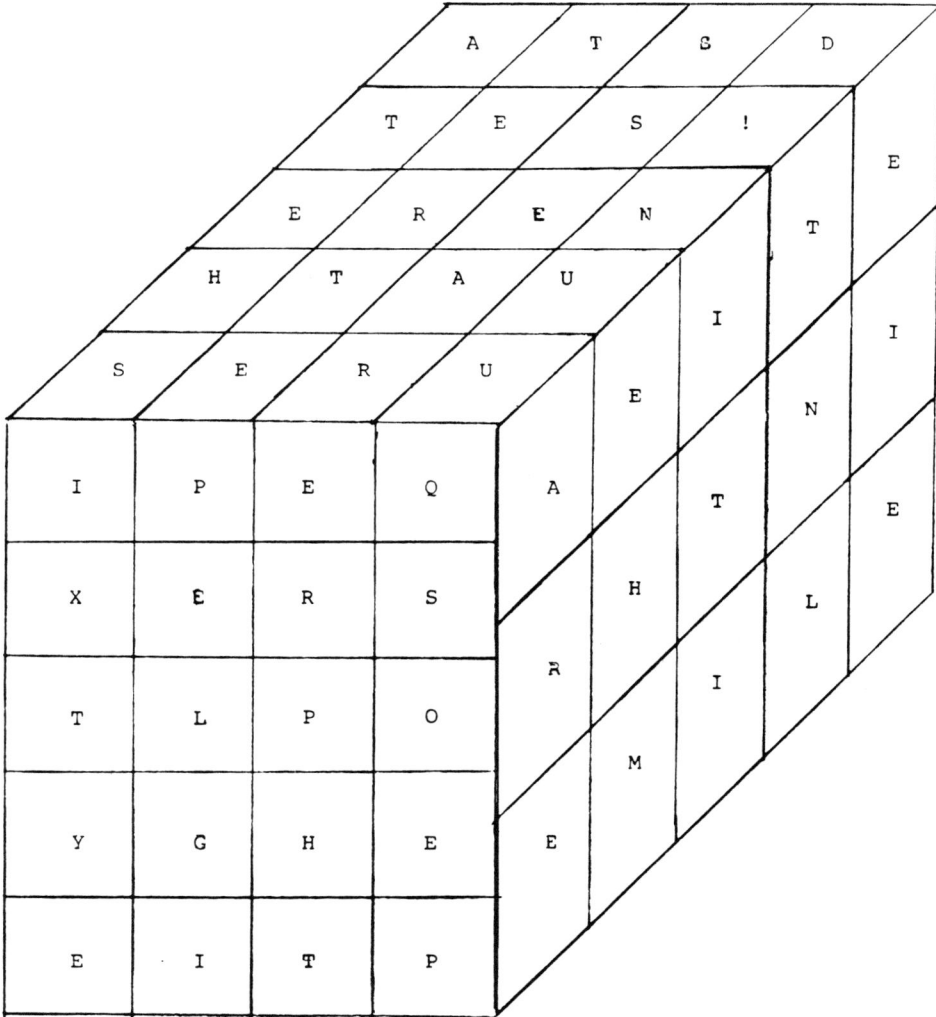

An Alarming Rate

Sentence Search

An Alarming Rate—Moving from one square to another up, down or sideways you'll learn something about children in this country. Every letter will be used once.

S	E	N	B	O	R	N	E
U	R	A	B	O	U	T	V
E	D	E	T	H	E	E	E
H	L	R	A	E	R	I	R
T	I	H	C	T	H	G	Y
N	I	E	T	U	N	I	M

Rules Are Rules

Sentence Search

Rules are rules—Moving from one square to another up, down or sideways a very important rule will be disclosed. Every letter will be used once.

V	E	A	B	A	S	E
E	R	N	I	I	A	P
R	O	D	N	R	I	N
Y	F	A	S	R	E	G
T	A	N	I	E	H	T

What part of the rule is missing?

Sniff, Sniff . . . Ahh Choooo!!!!

Instructions: Oops . . . someone sneezed and blew all the squares apart from one another. However, not all is lost; for there are microfilaments and microtubules holding the squares together in their original positions. The microfilaments (solid lines) hold the squares together horizontally while the microtubules (broken lines) hold the squares together vertically. Move from one square to another (continuously) vertically, horizontally and diagonally and an interesting piece of information will be disclosed.

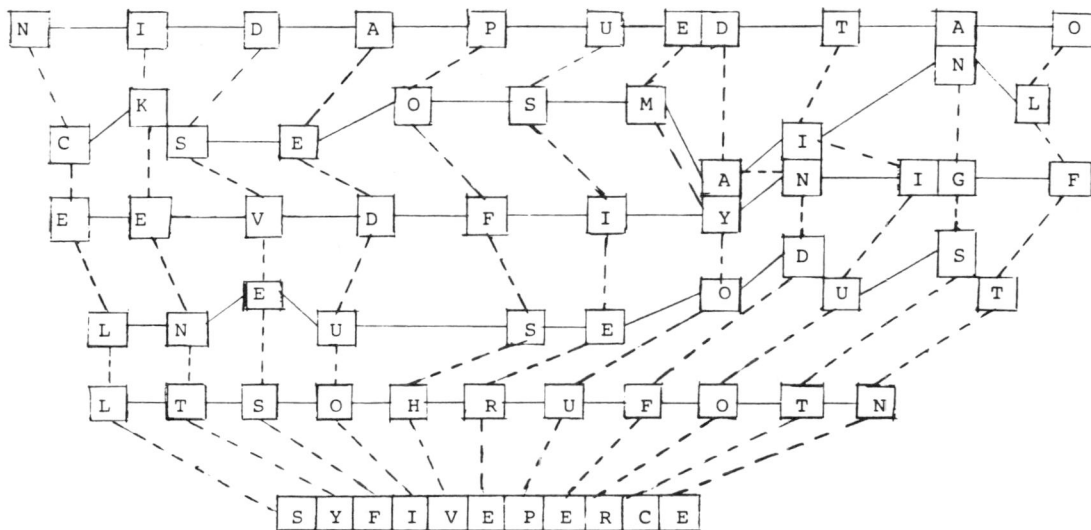

PART FIVE

Chemistry for Fun

A special series of non-threatening games/questions require that you use the periodic chart of the elements as a means of answering questions presented in some of the activities here. Other activities require your time and patience (for teachers of science: biological questions may also be incorporated in this scheme using the periodic table of the elements).

The Periodic Table of the Elements

	I-A	II-A	III-B	IV-B	V-B	VI-B	VII-B	VIII-B			I-B	II-B	III-A	IV-A	V-A	VI-A	VII-A	VIII-A
First Period	1 H 1.0080					Metals									Nonmetals			2 He 4.003
Second Period	3 Li 6.940	4 Be 9.013											5 B 10.82	6 C 12.010	7 N 14.008	8 O 16.0000	9 F 19.00	10 Ne 20.183
Third Period	11 Na 22.997	12 Mg 24.32				Transition Elements							13 Al 26.98	14 Si 28.09	15 P 30.975	16 S 32.066	17 Cl 35.457	18 Ar 39.944
Fourth Period	19 K 39.100	20 Ca 40.08	21 Sc 44.96	22 Ti 47.90	23 V 50.95	24 Cr 52.01	25 Mn 54.94	26 Fe 55.85	27 Co 58.94	28 Ni 58.71	29 Cu 63.54	30 Zn 65.38	31 Ga 69.72	32 Ge 72.60	33 As 74.92	34 Se 78.96	35 Br 79.916	36 Kr 83.80
Fifth Period	37 Rb 85.48	38 Sr 87.63	39 Y 88.91	40 Zr 91.22	41 Nb 92.91	42 Mo 95.95	43 Tc [99]	44 Ru 101.1	45 Rh 102.91	46 Pd 106.4	47 Ag 107.880	48 Cd 112.41	49 In 114.82	50 Sn 118.70	51 Sb 121.76	52 Te 127.61	53 I 126.91	54 Xe 131.3
Sixth Period	55 Cs 132.91	56 Ba 137.36	57—71 Lantha-nide Elements	72 Hf 178.50	73 Ta 180.95	74 W 183.86	75 Re 186.32	76 Os 190.2	77 Ir 192.2	78 Pt 195.23	79 Au 197.0	80 Hg 200.61	81 Tl 204.39	82 Pb 207.21	83 Bi 209.00	84 Po [210]	85 At [210]	86 Rn 222
Seventh Period	87 Fr [223]	88 Ra [226]	89- Actinide Elements															

Lanthanide (Rare Earth) Elements	57 La 138.92	58 Ce 140.13	59 Pr 140.91	60 Nd 144.27	61 Pm [147]	62 Sm 150.35	63 Eu 152.0	64 Gd 157.26	65 Tb 158.93	66 Dy 162.46	67 Ho 164.94	68 Er 167.27	69 Tm 168.94	70 Yb 173.04	71 Lu 174.99
Actinide Elements	89 Ac [227]	90 Th 232.05	91 Pa [231]	92 U 238.07	93 Np [237]	94 Pu [242]	95 Am [243]	96 Cm [251]	97 Bk [249]	98 Cf [251]	99 Es [254]	100 Fm [253]	101 Md [256]	102 No [253]	

An atomic weight value in brackets indicates the mass of the most stable known isotope.

Solving the Chemistry Problem

Instructions: Use the periodic table of the elements and write the chemical symbols of each of the elements being described (either by name, characteristics, positions on the periodic table, etc.). By combining all the chemical symbols together a well known definition in chemistry will be revealed to you. Oh yes, in some instances there is no information given about the element in which case you must determine the missing symbols yourself. In the immortal words of the great detective Sherlock Holmes, 'It's elementary my dear Watson.''

1. contains 90 electrons
2. no clue
3. atomic number 85
4. gas that sustains life
5. no clue
6. weighs 127 amu
7. has 16 neutrons
8. weighs 232 amu

9. no clue
10. found between Pm/Eu
11. Column IIIA, Period 3
12. no clue
13. $E = mc^2$
14. no clue
15. has 1 more proton than Bi
16. no clue
17. Titanium
18. contains 8 electrons
19. highest award in science
20. weighs 19 amu
21. no clue
22. an inert gas that begins and ends with the same letter
23. no clue (4x)
24. Nitrogen

25. Thorium
26. contains 85 electrons
27. atomic number 16
28. column IVB, row 4
29. no clue
30. named after Dr. Lawrence
31. no clue
32. its atomic number is the same as the mass of Ge
33. atomic number 49
34. contains 16 electrons
35. no clue
36. second element in the period table
37. rare earth in column VIB
38. has 8 neutrons
39. has a mass of 31 amu
40. Erbium
41. mass is twice that of Mg
42. atomic number 99
43. Oxygen
44. column VII, row 2
45. no clue
46. contains 2 electrons
47. no clue (5x)
48. found between carbon and nitrogen
49. no clue

84

Chemistry Done Periodically

Instructions: Using the periodic table of the elements given on the previous page answer the questions below:

> **Example**—What is the symbol for the element found in column IA and period 2? **Answer:** Li

1. I am a compound made up of atoms from column IA and VIIA. In column IA I have 3 electrons in the outer shell. In column VIIA I am composed of 10 neutrons. Who am I?

2. I'm positive and negative but never in between. What am I?

3. Take the following compound and oxidize it. What am I?

   CHOHLHUHMHNHIIIHAHRHOHWH2H

4. The sum of barium and the square of sodium will be very tasty. What is it?

5. In math, the sum of four zeros is zero. In chemistry, the sum of four zeros is 64 or 760. How can that be?

6. A person had four containers filled with the following substances: iodine, sodium, hydrogen, and carbon. What country is the person from?

7. This favorite dessert is found in the word America except that Americium must be reversed, the letters ''a & i'' must be removed, and the resulting answer must be placed backwards in ice.

8. An element from column IA when combined to another element in column VIIA has a total a.m.u. equal to the number of protons found in cesium. Identify the two elements.

9. Which element has an an* that is twice that of boron and half that of _____? Answer the question as well as completing the blank.

10. What do the following elements have in common?

    carbon   boron   oxygen
    argon   nitrogen

11. A wife gave a note to her absentminded husband:

    bromine, iodine, nitrogen, gallium, polonium, uranium
    neodymium, hydrogen, americium

    What was the message?

*an = atomic number

12. How many different elements can you find in this sentence?
13. Complete the missing sequence:

    $$1 + 1 = 4 \qquad 1 + 2 = 7 \qquad 1 + 3 = ? \qquad 1 + 4 = ?$$

14. Which element does not belong in this group? carbon, nitrogen, oxygen, helium, boron
15. Take the weight of water and subtract 3 dozen from it and you'll find out where the author went to school.
16. What movie and/or stage play does the following equation below represent?

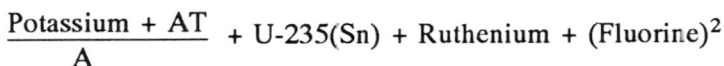

$$\frac{\text{Potassium} + \text{AT}}{\text{A}} + \text{U-235(Sn)} + \text{Ruthenium} + (\text{Fluorine})^2$$

Chemistry Done with Cross Words!

Complete the puzzle given the clues below. Unfortunately, the numbers were left out (puzzle and clues) so some letters have been inserted in the puzzle on page 87 to help you along.

Across

most common example of a monosaccharide

inorganic part of DNA

the number usually associated with atoms interacting

simplest form of matter

a process involved where hydrogen is removed from a molecule

atomic number of nitrogen

the opposite of oxidized

chemical symbol of the atom with an atomic number of 66

one of the most important atoms in organic compounds

the number of electrons found in hydrogen

the first atom listed on the periodic chart

a nucleic acid

the central mass of the atom

what a hen lays

Down

another term for fats

a meal taken at the end of the day

negative particles found in the atom

positively charged particles of the atom

opposite of negative

holds your teeth in place

type of bond in proteins

type of bond where atoms give away or receive electrons

a general class or organic compounds usually referred to as sugars

a nucleic acid

the charge carried by an electron

an atom with chemical symbol, N

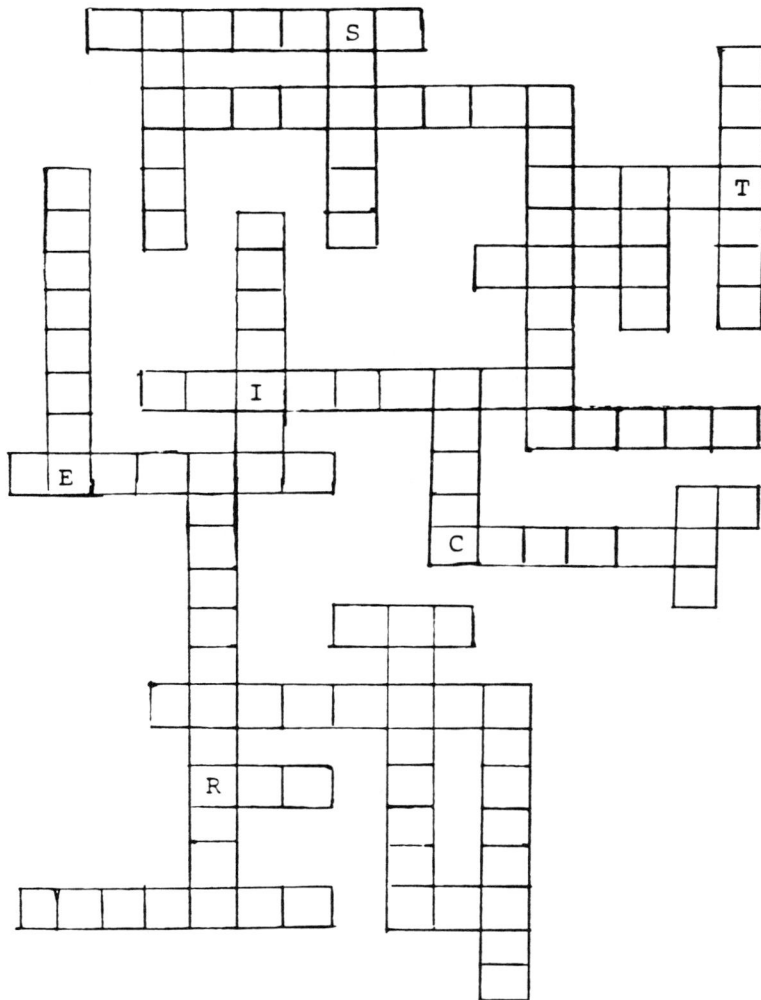

On Making an Organic Compound

Instructions: Link all of the carbon atoms with as few lines* as possible being careful not to join the nitrogen atom. Be sure that all carbon atoms are attached to one another.

```
C       C       C       C       C

C       C       N       C       C

C       C       C       C       C

C       C       C       C       C

C       C       C       C       C
```

*Eight is the magic combination. All lines are to be interconnected.

On Splitting the Hydrogen Atoms

Twenty-four hydrogen atoms are closely held together by their strong attraction towards one another. However, it's time to "split" them into two equal groups (12 hydrogen atoms per group) before they get out of hand. To make sure all is well split the atoms apart so that they form symmetrical groups.

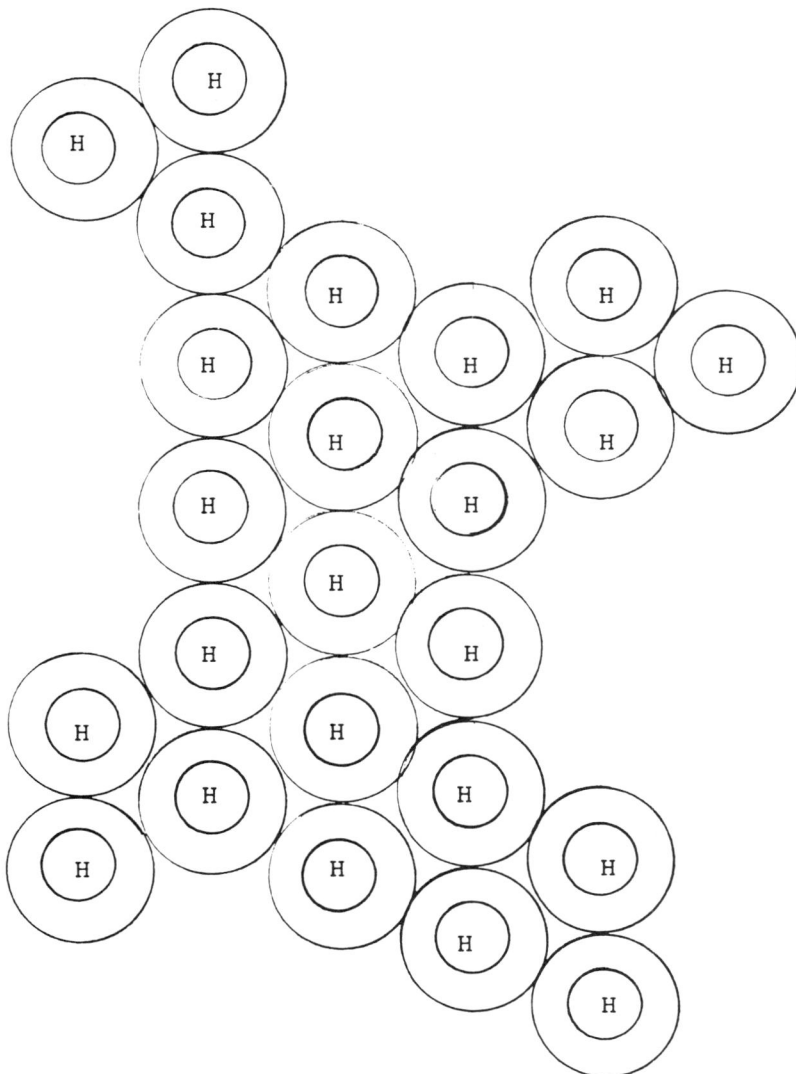

PART SIX

Fill-In and/or Color

It is highly recommended that you use color pencils in the following activities. A good reference book (general biology) would also be very helpful. The activities presented here have two functions: One, you are asked to fill in the missing letters to complete words that are partially spelled out. Two, you are asked to select a color (your choice) and to color the circle and its corresponding structure. In this instance, if you wish to determine if your answers are correct you must use the reference book. The answer sheet in the back of the book only provides answers to the word completion portion of the activity (of course coloring of the structures are optional but it represents an excellent means of learning a subject matter).

Generalize Plant Cell

Fill in and/or Color

Instructions: Fill in the missing letters to complete the names of the structures belonging to the plant. Create your own legend by coloring each circle and its corresponding structures. A general biology text would make an excellent reference to check your answers.

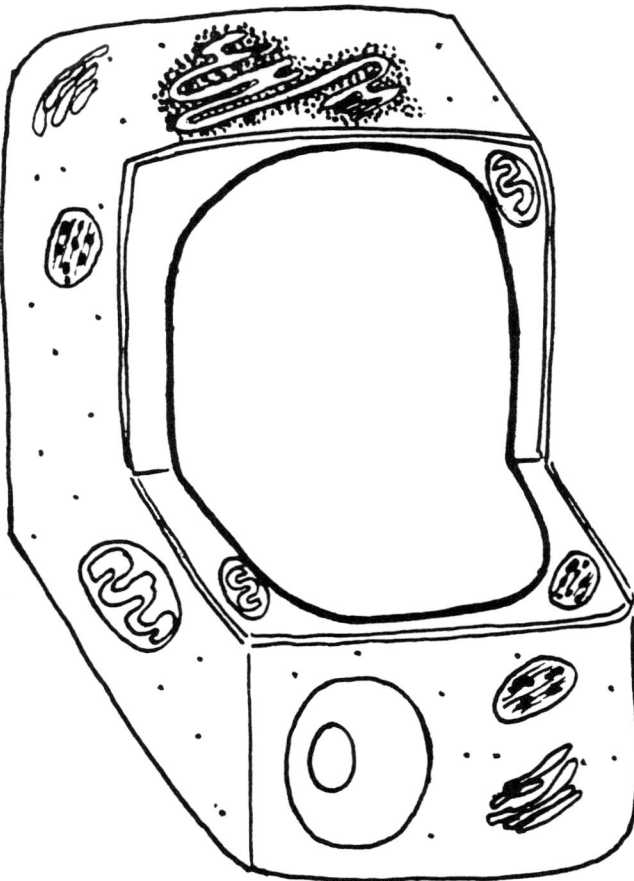

○ P_AS_A _EM_RA_E

○ _LAS_OD_SM_T_

○ C_LO_OP_A_T

○ _ACU_L_

○ _IT_CHO_DR_ON

○ R_BO_OM_

○ _NDO_LA_MI_
   R_TI_UL_M

○ G_L_I _P_AR_TU_

○ _U_LE_S

○ N_C_EO_U_

○ C_RO_A_I_

○ NU_L_A_
   _E_BR_N_

○ _UC_E_R _A_

Typical Flower

Fill in and/or Color

Instructions: Fill in the missing letters to complete the names of the structures belonging to the plant. Create your own legend by coloring each circle and its corresponding structures. A general biology text would make an excellent reference to check your answers.

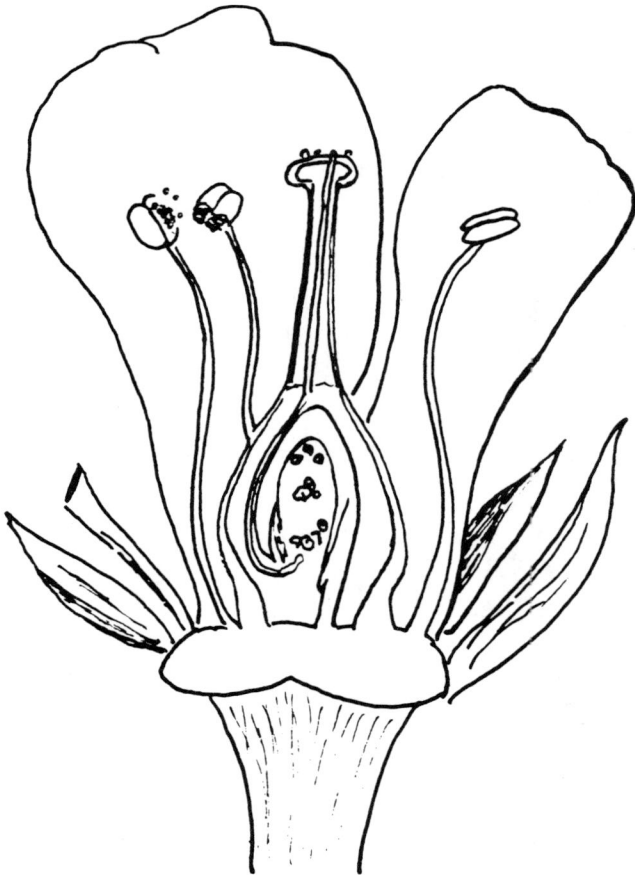

O __NT__E__
O F_L_ME_T   __TA__EN

O __TI_M__
O S_Y_E
O O_A_Y   __IS__I__
O __VU_E__

O P_TA__
O __E_A__
O P_D_NC_E
O __EC_PT_C_E

94

Structure of a Seed (Corn)

Fill in and/or Color

Instructions: Fill in the missing letters to complete the names of the structures belonging to the plant. Create your own legend by coloring each circle and its corresponding structures. A general biology text would make an excellent reference to check your answers.

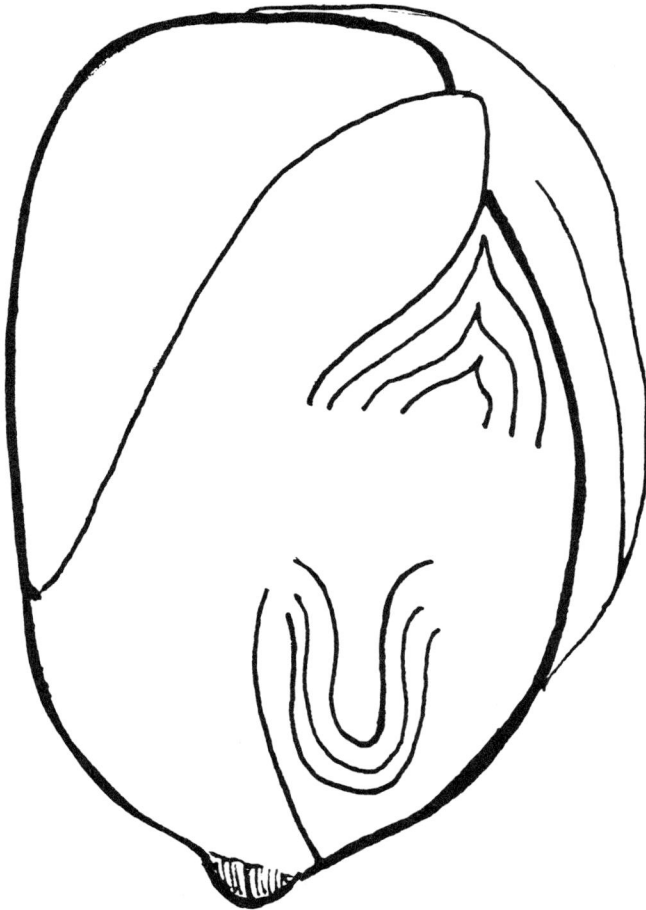

○ _US_D P_RI_AR_ AND _E_D _O_T

○ _ND_S__ER_

○ C_T_L_DO_

○ _O_EO_TI_E

○ E_IC_T_L ⎤ __MB_Y_

○ _YP_C_TY_

○ _AD_C_E ⎦

○ _OL_O__HI_A

○ _ED_NC_E

Simple Complete Leaf

Fill in and/or Color

Instructions: Fill in the missing letters to complete the names of the structures belonging to the plant. Create your own legend by coloring each circle and its corresponding structures. A general biology text would make an excellent reference to check your answers.

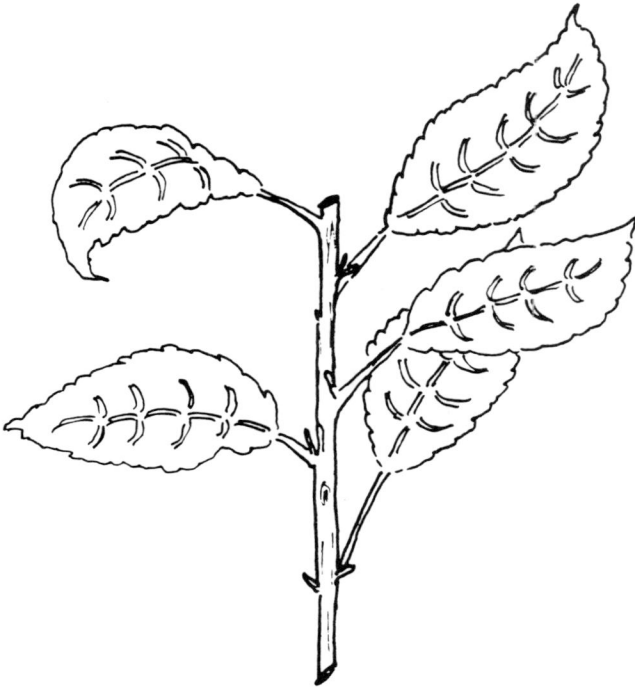

○ __LE__DE

○ P__TI__L__

○ __TI__UL__S

○ __X__L

○ __NT__RN__D__

○ __O__E

Cross Section of a Leaf

Fill in and/or Color

Instructions: Fill in the missing letters to complete the names of the structures belonging to the plant. Create your own legend by coloring each circle and its corresponding structures. A general biology text would make an excellent reference to check your answers.

○ C_TI_L_

○ _PP_R _PI_ER_I_

○ _YL_M_ ⎤
⎟ _E_N
○ _LO_M ⎦

○ A_R _P_C_

○ P_LI_A_E ⎤
⎟ _ES_ PH_L_
○ S_ON_Y ⎦

○ _UA_D _E_L

○ S_OM_T_

○ L_W_R E_ID_RM_S

Bacteria Cell

Fill in and/or Color

Instructions: Fill in the missing letters to complete the names of the structures belonging to the bacteria. Create your own legend by coloring each circle and its corresponding structures. A general biology text would make an excellent reference to check your answers.

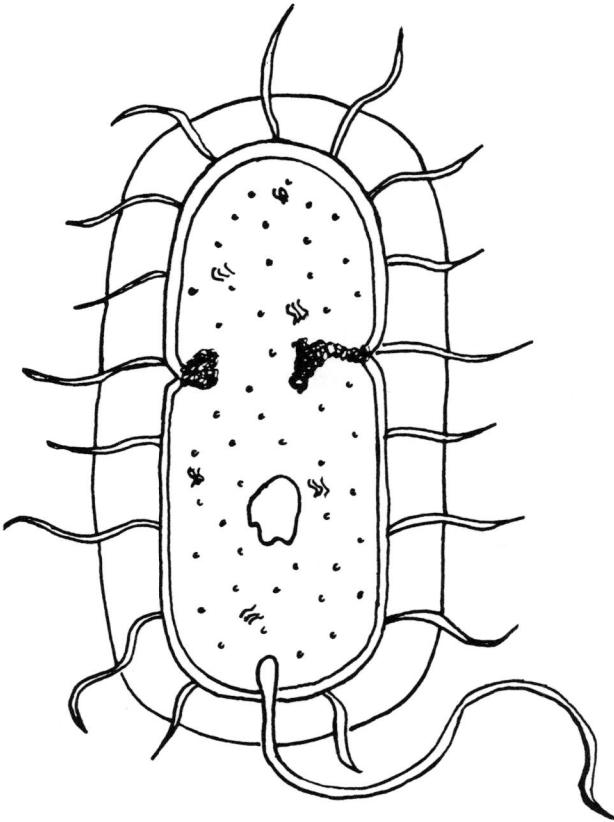

○ F_MB_I_

○ _LA_EL_U_

○ C_PS_L_

○ _E_L_A_L

○ C_TO_LA_MI_ _EM_RA_E

○ _UC_E_R __EM_R_N_

○ M_S_ME_

○ _NC_ U_I_N _RA_U_E_

○ R_B_SO_E_

○ _E_T_M

Generalize Animal Cell

Fill in and/or Color

Instructions: Fill in the missing letters to complete the names of the structures belonging to the animal. Create your own legend by coloring each circle and its corresponding structures. A general biology text would make an excellent reference to check your answers.

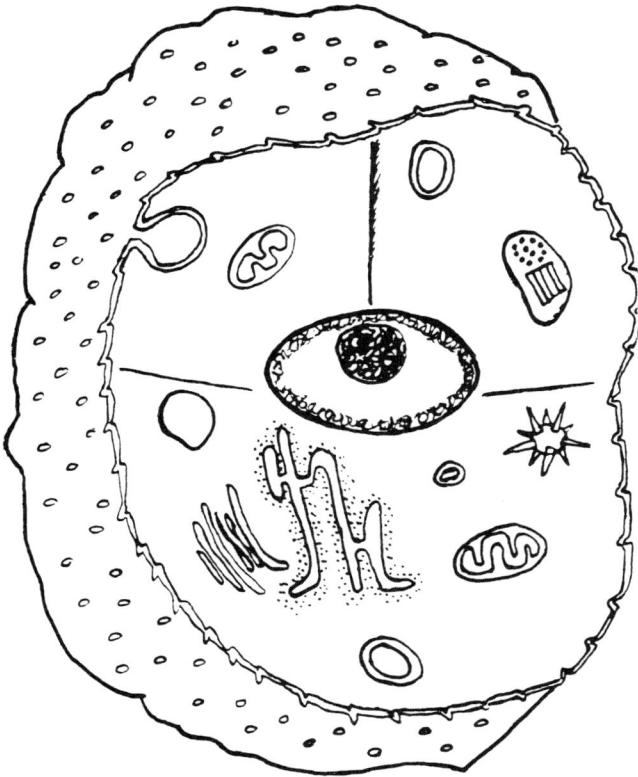

○ _E_ L _ _E_ BR_ N_

○ C_ T_ PL_ S_

○ _IB_ S_ ME_

○ M_ T_ CH_ ND_ I_ N

○ _UC_ E_ S

○ C_ R_ MA_ I_

○ N_ C_ EO_ U_

○ _EN_ _ _I_ L_ S

○ E_ D_ PL_ S_ I_
   RE_ I_ U_ U_

○ _OL_ I_ _P_ RA_ US

○ L_ S_ SO_ E_

○ _I_ O_ YT_ C_ _ES_ C_ E

99

The Ameba

Fill in and/or Color

Instructions: Fill in the missing letters to complete the names of the structures belonging to the animal. Create your own legend by coloring each circle and its corresponding structures. A general biology text would make an excellent reference to check your answers.

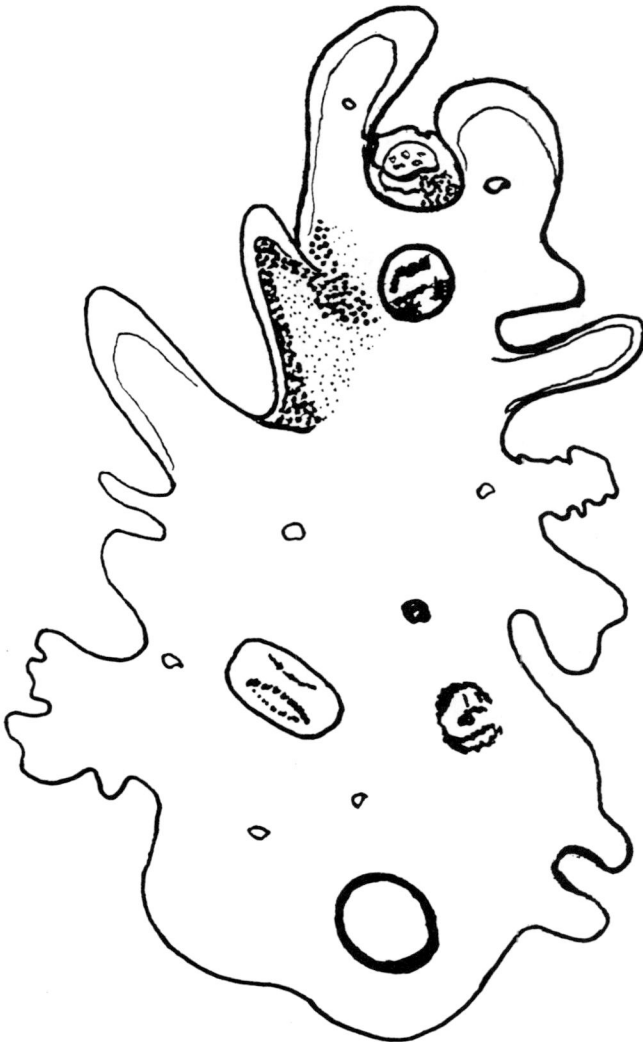

○ _ U _ L _ U _

○ C _ NT _ AC _ IL _ V _ CU _ L _

○ _ OO _ _ AC _ O _ E

○ A _ A _ _ OR _

○ PL _ S _ AL _ M _ A

○ H _ A _ I _ E _ A _

○ P _ AS _ AG _ L

○ _ LA _ MA _ O _

○ _ SE _ D _ P _ D

The Paramecium

Fill in and/or Color

Instructions: Fill in the missing letters to complete the names of the structures belonging to the animal. Create your own legend by coloring each circle and its corresponding structures. A general biology text would make an excellent reference to check your answers.

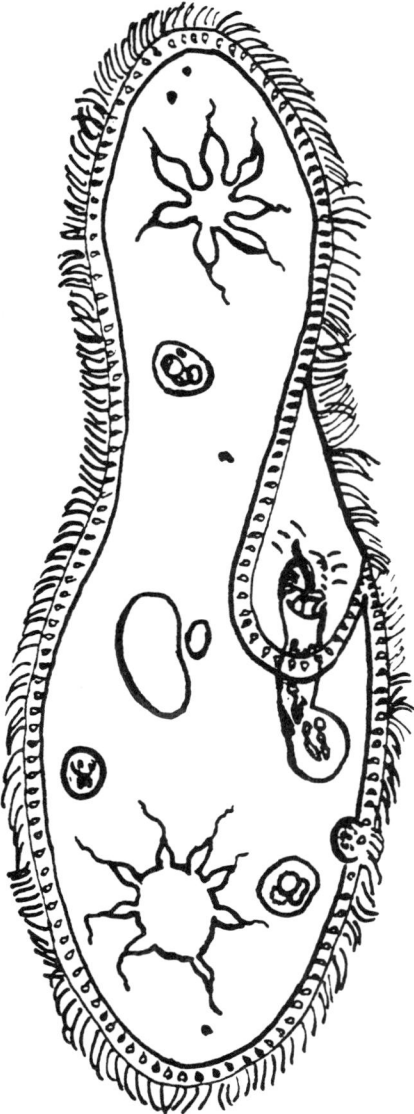

○ _ I _ I _

○ P _ LL _ C _ E

○ _ _ CT _ PL _ S _

○ E _ D _ P _ A _ M

○ _ R _ CH _ C _ S _ S

○ O _ A _ _ RO _ V _

○ MA _ R _ NU _ L _ U _

○ _ IC _ ON _ C _ E _ S

○ C _ TO _ T _ M _

○ _ Y _ OP _ A _ YN _

○ FO _ D _ _ A _ UO _ E

○ _ N _ L _ _ OR _

○ CO _ T _ AC _ I _ E
      V _ C _ O _ E

○ _ AD _ A _ I _ G _ _ A _ A _

The Hydra

Fill in and/or Color

Instructions: Fill in the missing letters to complete the names of the structures belonging to the animal. Create your own legend by coloring each circle and its corresponding structures. A general biology text would make an excellent reference to check your answers.

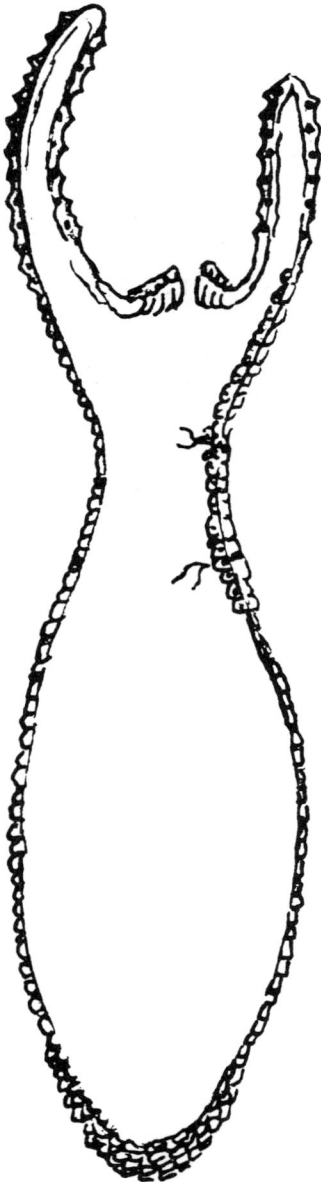

○ _ O _ T _

○ _ Y _ OS _ O _ E

○ C _ I _ OB _ A _ T

○ _ EN _ AC _ E

○ E _ ID _ R _ I _

○ _ ES _ G _ E _

○ G _ ST _ OD _ R _ I _

○ _ AS _ RO _ AS _ UL _ _ CA _ I _ Y

○ F _ A _ EL _ U _

The Nerve Tissue

Fill in and/or Color

Instructions: Fill in the missing letters to complete the names of the structures belonging to the animal. Create your own legend by coloring each circle and its corresponding structures. A general biology text would make an excellent reference to check your answers.

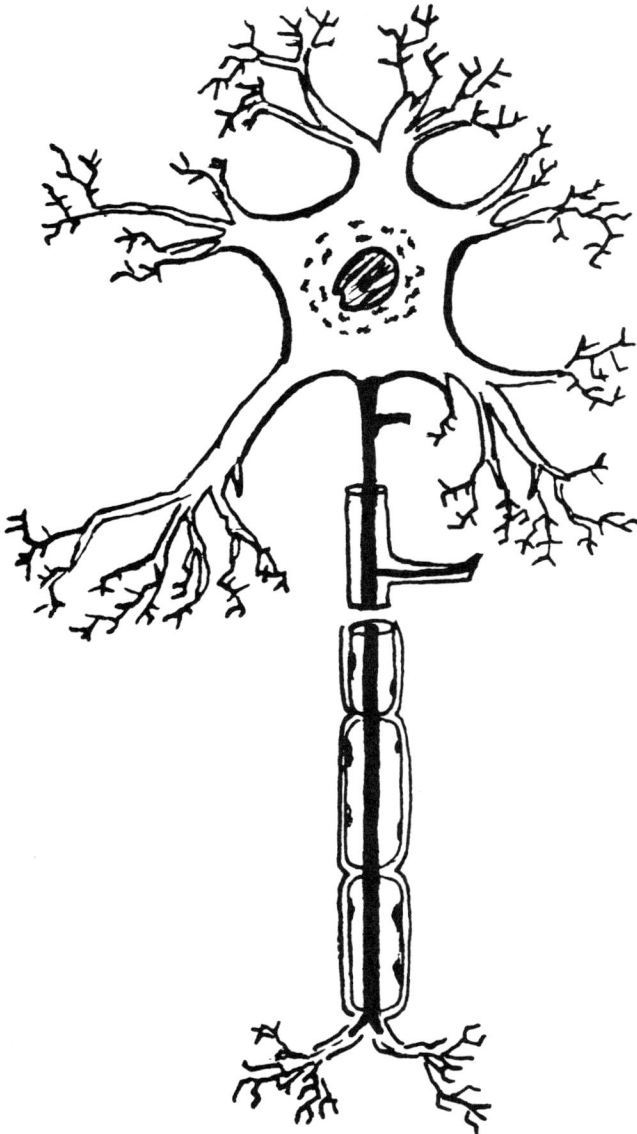

O __EN__R__T__

O N__RV__ __E__L __O__Y

O __US__E__S

O N__C__EO__U__

O __X__N

O __YE__I__ __HE__T__

O __OD__ __F __AN__I__R

O __UC__E__S O__
__CH__A__M __E__L

The Human Skull

Fill in and/or Color

Instructions: Fill in the missing letters to complete the names of the structures belonging to the organ system. Create your own legend by coloring each circle and its corresponding structures. A general biology text would make an excellent reference to check your answers.

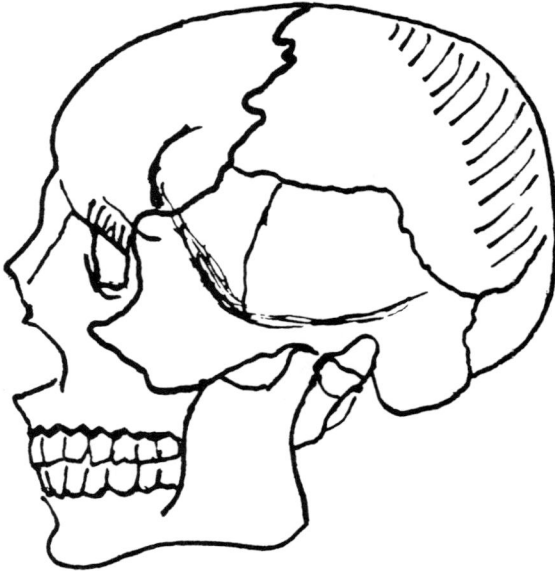

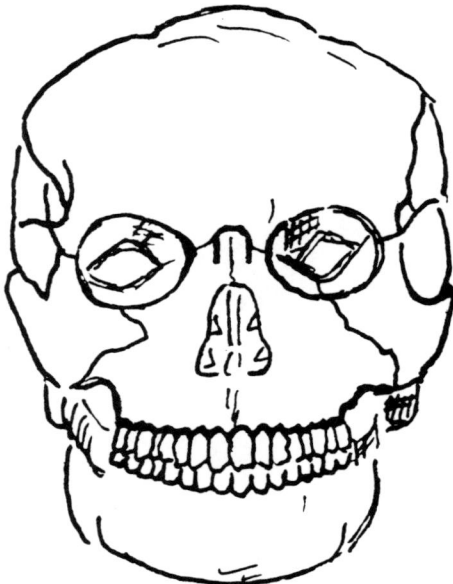

Cranial Bones

O _ C _ IP _ T _ L

O _ A _ IE _ A _

O F _ O _ T _ L

O _ E _ P _ RA _

O E _ HM _ I _

O _ P _ E _ OI _

Facial Bones

O _ AS _ L

O _ OM _ R

O L _ C _ IM _ L

O _ YG _ MA _ I _

O P _ L _ TI _ E

O _ A _ I _ L _

O MA _ D _ B _ E

O I _ F _ RI _ R _ _ AS _ L C _ ON _ HA

The Bronchial Tree

Fill in and/or Color

Instructions: Fill in the missing letters to complete the names of the structures belonging to the organ system. Create your own legend by coloring each circle and its corresponding structures. A general biology text would make an excellent reference to check your answers.

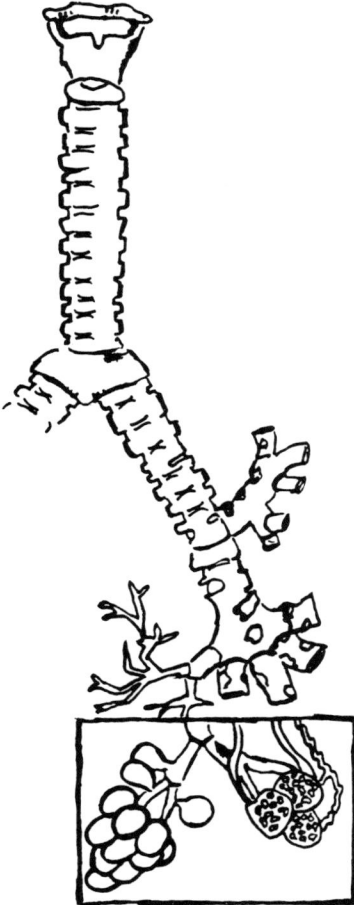

○ _ YO _ D   B _ NE

○ T _ Y _ OI _   C _ R _ IL _ G _

○ _ R _ CO _ D   _ ART _ LA _ E

○ _ RA _ H _ A

○ B _ O _ CH _ S

○ _ R _ N _ H _ OL _

○ R _ S _ IR _ TO _ Y   BR _ N _ HI _ L _

○ _ LV _ O _ A _   _ U _ T

○ A _ VE _ LA _   _ A _

The Human Heart

Fill in and/or Color

Instructions: Fill in the missing letters to complete the names of the structures belonging to the organ system. Create your own legend by coloring each circle and its corresponding structures. A general biology text would make an excellent reference to check your answers.

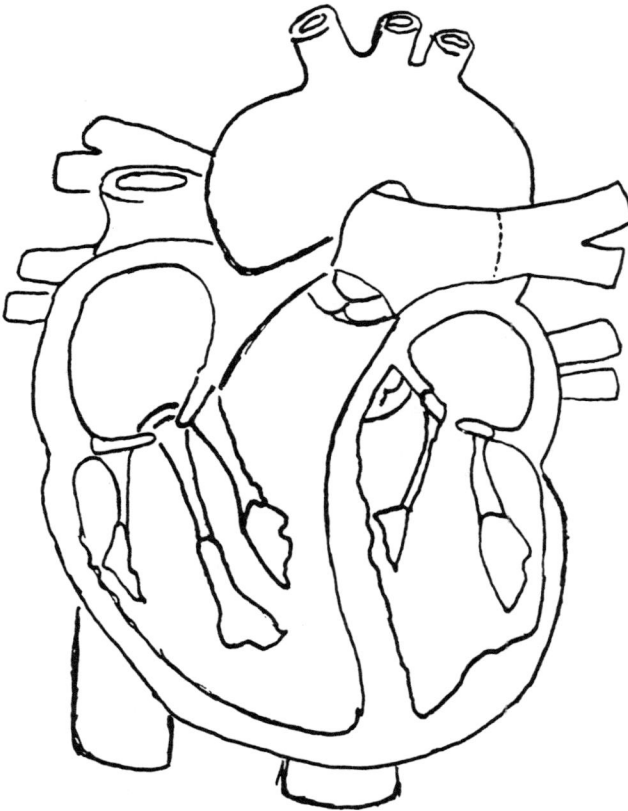

○ _ O _ T _

○ A _ RT _ C _ EM _ L _ NA _ VAL _ E

○ _ I _ HT _ TR _ U _

○ T _ IC _ SP _ D _

○ R _ GH _ _ E _ TR _ C _ E

○ _ UL _ ON _ R _ S _ MI _ U _ A _ _ AL _ E

○ P _ LM _ NA _ Y _ E _ N

○ L _ F _ A _ RI _ M

○ _ IC _ S _ ID (_ IT _ AL V _ L _ E)

○ L _ F _ _ E _ T _ I _ L _

○ S _ P _ RI _ R V _ N _ C _ V _

○ _ NF _ RI _ R V _ N _ CA _ A

○ C _ O _ DA _ T _ ND _ NE _ E

○ _ AP _ L _ AR _ _ U _ C _ E

The Digestive System

Fill in and/or Color

Instructions: Fill in the missing letters to complete the names of the structures belonging to the organ system. Create your own legend by coloring each circle and its corresponding structures. A general biology text would make an excellent reference to check your answers.

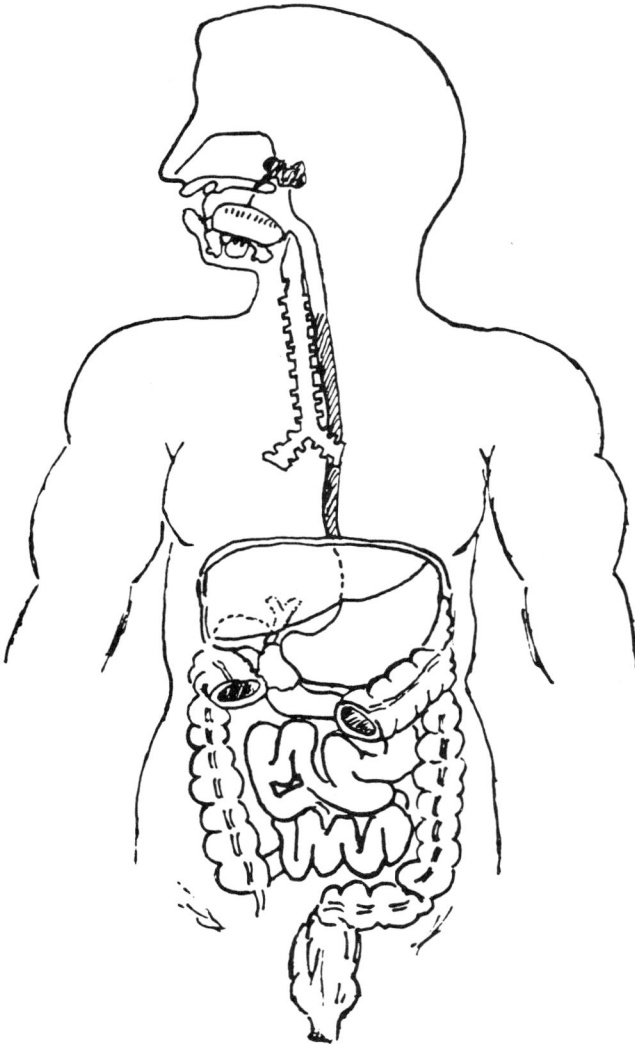

○ _RAL _AV_T_

○ P_AR_N_

○ S_OM_C_

○ E_OP_A_I_

○ _MA_L _NT_ST_NE_

○ L_R_E IN_ES_I_ES

○ _EC_M

○ _P_EN_I_

○ C_L_N

○ _EC_U_

○ A_A_ _AN_L

○ _E_T_

○ T_NG_E

○ _AL_V_RY _LA_D_

○ _I_E_

○ G_L_ _L_DD_R

○ _A_CR_A_

○ _P_E_N

The Human Kidney

Fill in and/or Color

Instructions: Fill in the missing letters to complete the names of the structures belonging to the organ system. Create your own legend by coloring each circle and its corresponding structures. A general biology text would make an excellent reference to check your answers.

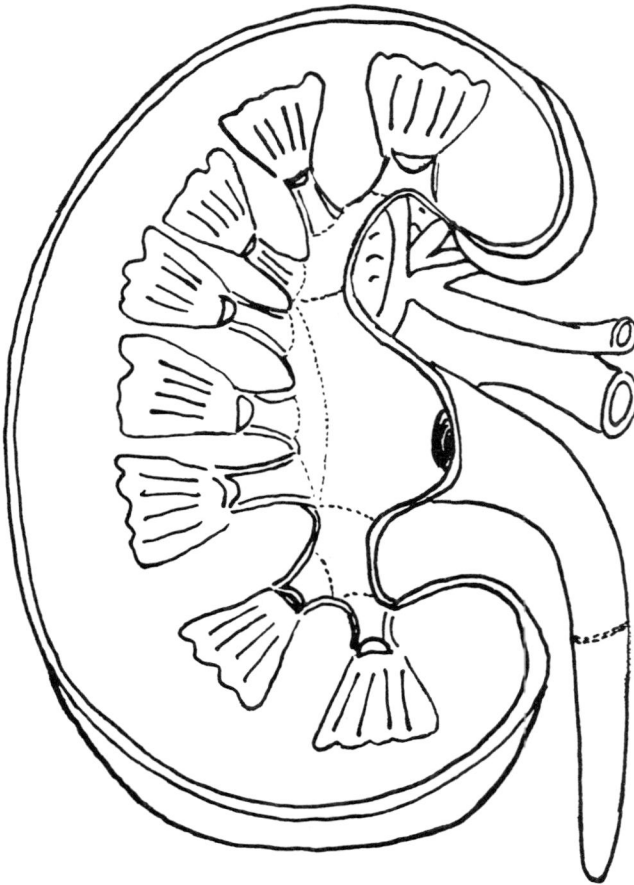

○ _EN_L _ _AP_UL_

○ C_RT_X_

○ _E_UL_A

○ P_P_LL_ _

○ _AL_X _ _I_O_

○ C_LY_ M_JO_

○ _E_A_ _EL_I_

○ _R_T_R

○ R_N_L A_T_R_ _

○ _EN_L _E_N

The Endocrine System

Fill in and/or Color

Instructions: Fill in the missing letters to complete the names of the structures belonging to the organ system. Create your own legend by coloring each circle and its corresponding structures. A general biology text would make an excellent reference to check your answers.

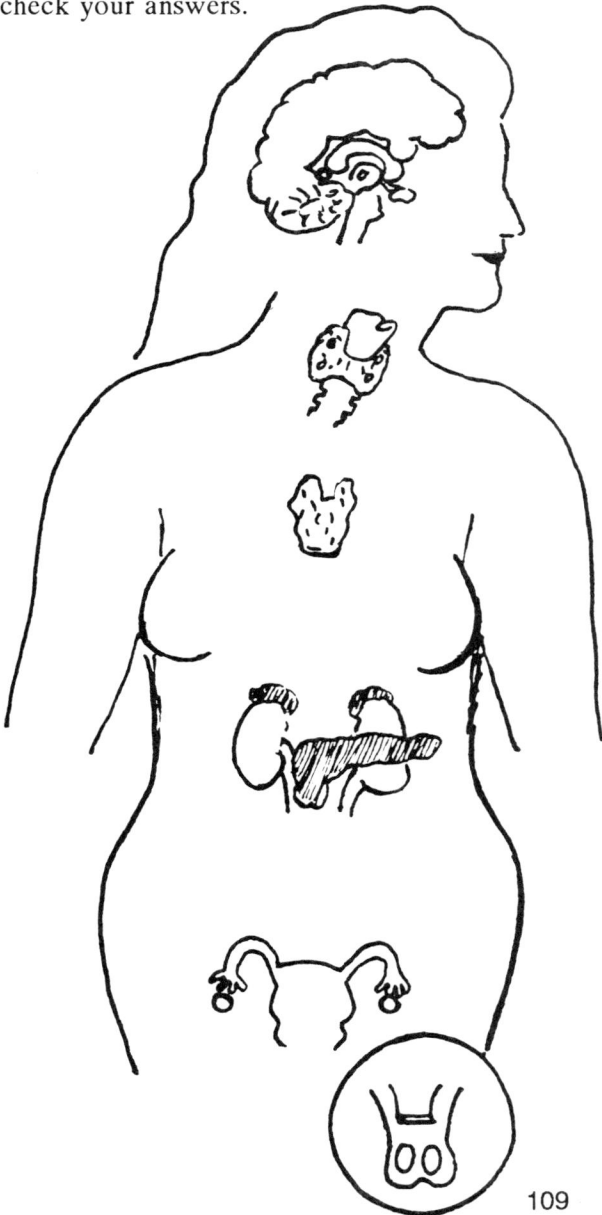

Glands

○ __ IT __ IT __ R __

○ __ I __ E __ L

○ T __ YR __ I __

○ __ AR __ TH __ R __ ID

○ __ H __ M __ S

○ A __ RE __ A __ S

○ __ A __ C __ EA __

○ __ V __ RI __ S

○ __ E __ TE __

Cut-Outs—Puzzles and Structures

What would an activities book be without having to cut out some things or structures? The next few pages allow one to explore not only cut-outs but also to construct biological structures as shown in the section called "Origami Biology." For a more complete enjoyment of "Origami Biology," coloring is highly recommended.

So Many Teases

Help! For some odd reason the thyroid glands had exploded into four pieces and that its reconstruction would result in a T-shaped rather than an H-shaped organ. As the physician in charge, please remove the four pieces (cut them out!) and reconstruct (like pieces of a puzzle) the T-shaped thyroid glands. Be careful of the T's within each piece; they may tease you.

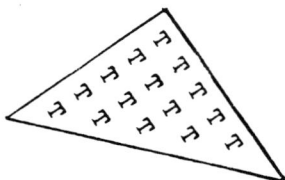

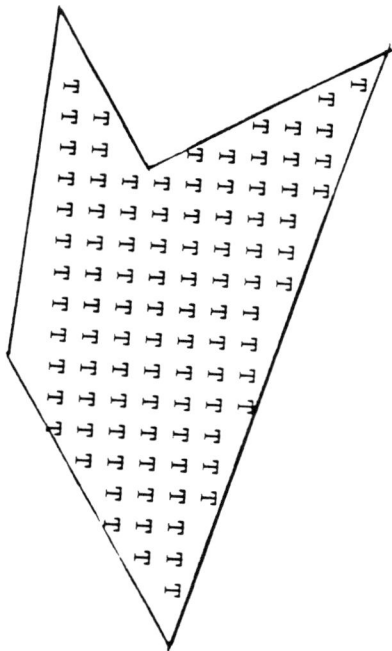

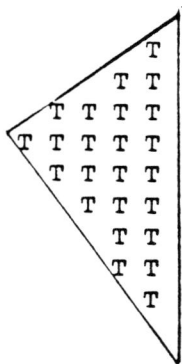

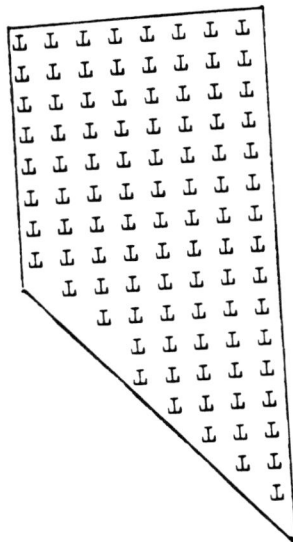

A Cut of Life

Instructions: Cut out the entire card and fold along the broken line. With the card open cut along solid lines. Fold letters outwards; all other parts are folded inward. See illustration below.

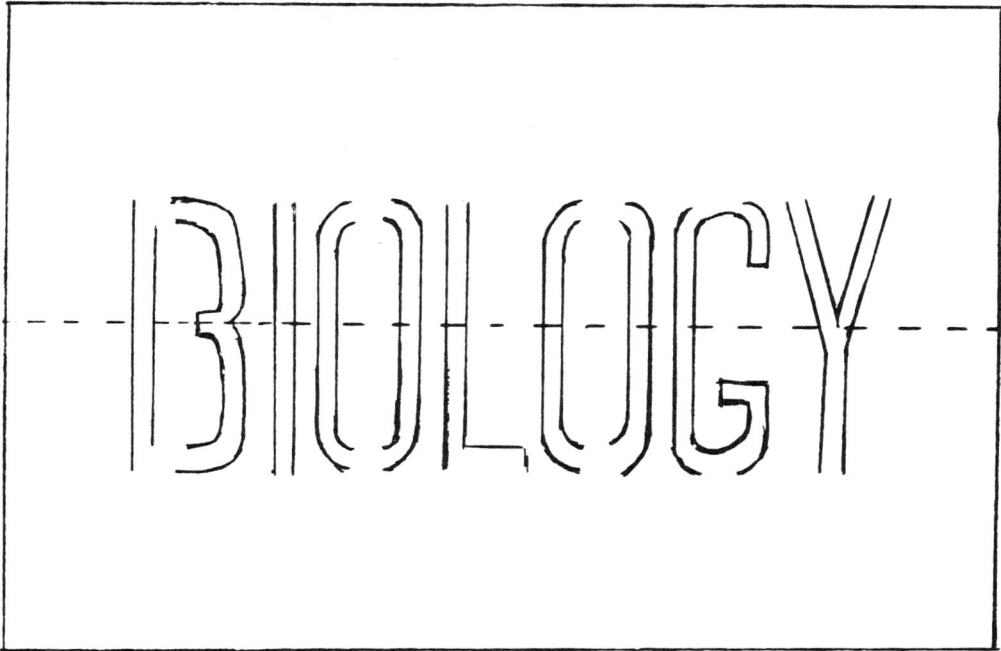

Note: It is highly suggested that you xerox this page and do a trial run. You may also wish to use different grades of paper. The complete alphabet and set of numbers are given on the following page for further exploration and fun. Enjoy!

ABCDEFGHIJK

LMNOPQRSTU

VWXYZ

12345678

Origami Biology—A Three Dimensional Approach
(Color/Cut/Construct)*

Perhaps one of the most atypical approaches used in the teaching and learning of biological material is "Origami Biology" (I really don't know if this term has ever been used). It is nothing more than paper folding of biological structures (that the reader may or may not wish to color) that will appear three-dimensional rather than in the two-dimensional plane as exhibited in most texts.

Origami Biology may really be a "laboratory" model that one could have at home for study, for pleasure or for both. Time, however, does not allow me to illustrate all of the models available in a laboratory setting. It is for this reason that I have provided a blank origami sheet so that you may create whatever structures you wish. Have Fun! (I would be more than happy to hear from you with regards to this approach to biology. Write c/o of Vic Chow, City College of San Francisco, San Francisco, CA. 94112)

*Reprinted with permission from Chow: Human Biology Lab Manual: A Matter of Survival. Copyright 1987 Kendall/Hunt Publishing Company.

Instructions for the Construction of the Origami Box*

(Note: It is suggested that you do a practice ''run'' with a plain sheet of paper)

1. Cut the paper from point A to point B (sides are equidistant)

2. Fold paper, point to point, from A to B' and from A' to B (open up the paper)

3. Fold paper, point to point, from A to B and from A' to B' (keep paper folded)

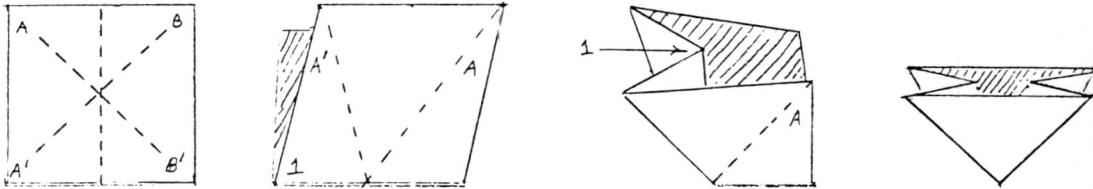

4. Turn paper sideways and fold paper at region 1 towards the center of the paper.

5. Repeat on other side and turn the paper-up-side down. (Shaded area represents back of the paper).

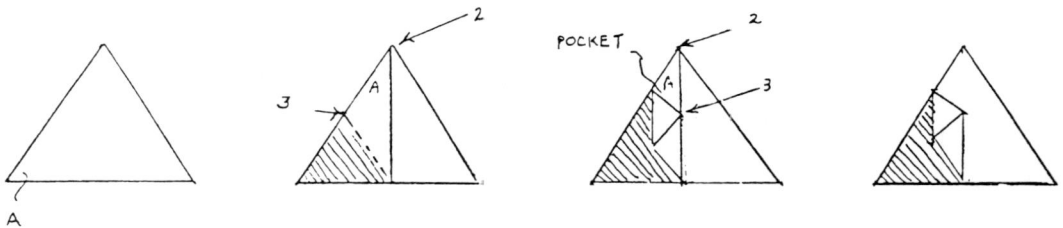

6. Fold tip of A towards region 2.

7. Fold tip of region 3 towards the center.

8. Place tip of region A into pocket of region 3.

9. Repeat procedures for tip of region A' on the right side.

*Reprinted with permission from Chow: Human Biology Lab Manual: A Matter of Survival. Copyright 1987 Kendall/Hunt Publishing Company.

10. Turn paper over and repeat steps 6–9 for regions B and B′.

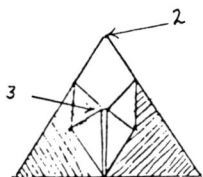

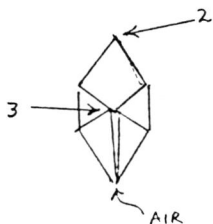

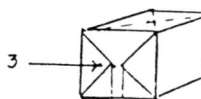

11. Blow into open region and your origami box is complete.

The outline below is a blank copy for duplicating (tracing or xeroxing). Some planning is advisable if you wish to draw your own biological structures in three-dimensions. Look over sample diagrams to get a general idea of how to orient your drawings. After you complete the diagram, color them in, cut and fold, and enjoy your own biological creations!

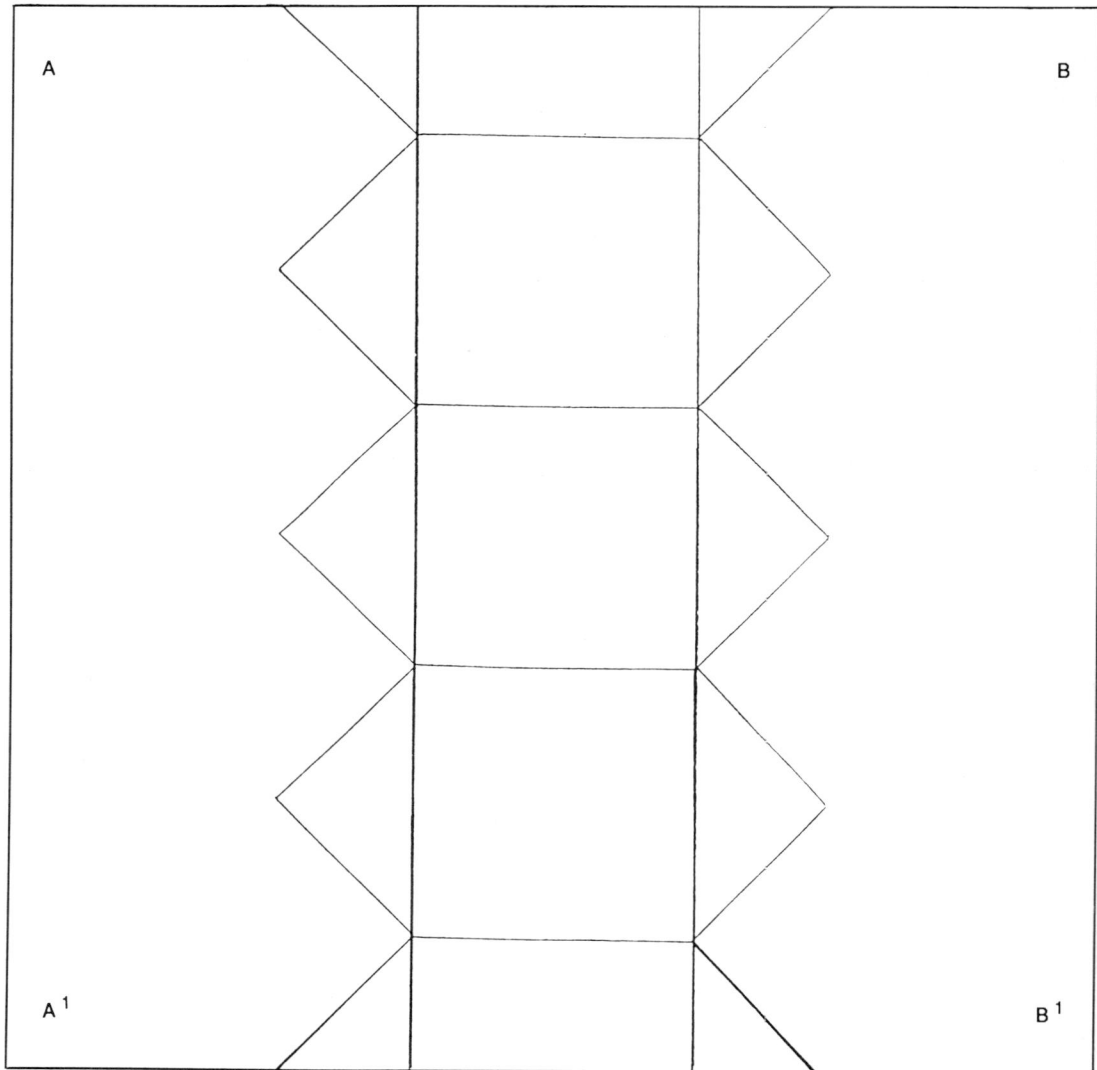

The Animal Cell*

Instructions: Identify the structures labelled A through L (center diagram—answers are printed on the back side). Color the structures in the center diagram. Now color in all the other cellular structures in the other "squares" (the triangles from point A to A′ and from B to B′ represent parts of the complete square). Construct the origami block as given in the instructions. Use your block for reviewing purposes. (Forget the names of the structures? Just peek through the opening of your block. The answers are inside!)

*Reprinted with permission from Chow: Human Biology Lab Manual: A Matter of Survival. Copyright 1987 Kendall/Hunt Publishing Company.

A. Nucleus
B. Chromosome
C. Rough E. R.
D. Smooth E. R.
E. Centriole
F. Vacuole
G. Lysosome
H. Microtubule
I. Golgi body
J. Mitochondrium
K. Cell membrane
L. Ribosome

The Compound Microscope*

Instructions: Identify the labelled structures of the microscope—answers are printed on the back side. Coloring of the microscope is optional. Construct the origami block and you will obtain a "true" 3-dimensional diagram of the compound microscope.

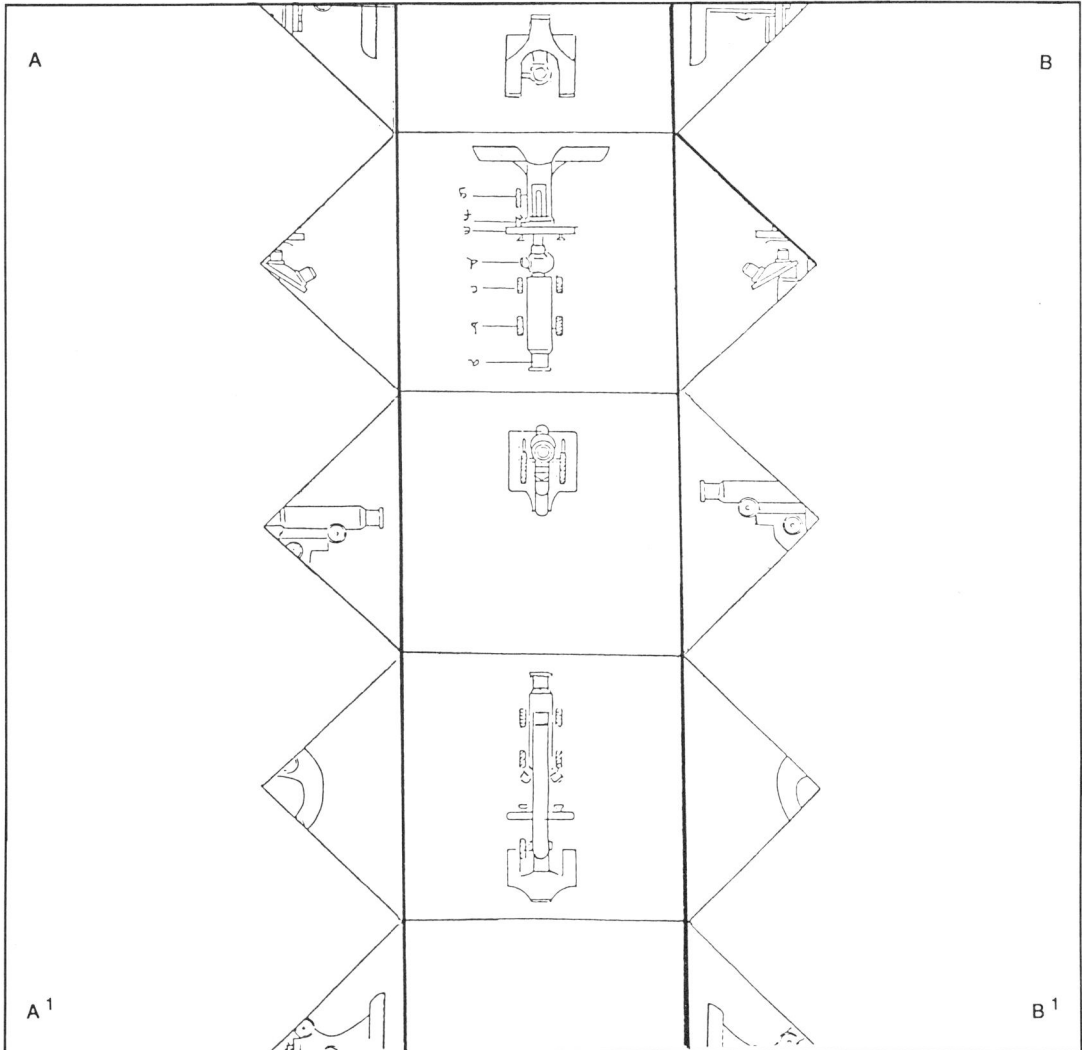

*Reprinted with permission from Chow: Human Biology Lab Manual: A Matter of Survival. Copyright 1987 Kendall/Hunt Publishing Company.

a. Eyepiece
b. Coarse adjusting knob
c. Fine adjusting knob
d. Objective lens
e. Stage
f. Condenser
g. Condenser knob

Mitosis in Action*

Instructions: Identify the labelled structures of the animal cell (answers are on the back). Each animal cell is shown in a different phase of mitosis. Can you identify the phases?—answers are on the back. Coloring is optional but is very pleasing to the eyes. Construct the origami block. For review, roll the block like a die and identify the mitotic phase that lands side up.

*Reprinted with permission from Chow: Human Biology Lab Manual: A Matter of Survival. Copyright 1987 Kendall/Hunt Publishing Company.

A. Interphase
B. Prophase
C. Anaphase
D. Metaphase
E. Telophase

The 3-Dimensional Puzzle

Construct the cube and with a felt pen try to work through the maze from the top (marked A,B) to the bottom (marked A′, B′). Obviously, it's easier to solve the maze on a 2-dimensional plane but try it on a 3-dimensional plane for fun!

PART EIGHT

Compounded Games

These are activities that will involve you having a partner to share in the enjoyment of learning biology. For example, in the activity entitled "Concentration," your partner can time you as well as ask you the questions (just to keep you honest—see following page). Other activities would require that you use a calculator, a pair of scissors, and color pencils. It is recommended that you "play" with these activities at home rather than at school, work, car, etc. Some time and a lot of patience are required. Do read the instructions to the activities carefully.

Concentration

This exercise illustrates the manner in which you pay attention to detail. Study the list of words below for the next two minutes. Pay particular attention to the spelling of the words. At the end of two minutes turn the page over and answer the questions.

SPONGES	GINKGOS	SQUID	FRUIT FLY
LIVERWORT	LOBSTER	HORSETAIL	PINE
EARTHWORM	REDWOOD	RABBIT	JUNIPER
SNAIL	HEMLOCK	ANEMONE	PALM
LARCH	SHARK	LILY	JELLYFISH

Questions

1. What was the title of this exercise?
2. What was the total number of sentences in this exercise?
3. How many times was the word "attention" used in this exercise?
4. How many columns of organisms were listed?
5. How many organisms were listed in each column?
6. What was the total number of organisms listed?
7. How many plants were listed? How many animals were listed?
8. What was the name of the plant and animal that ended with the letter H?
9. How many plants and animals ended with the same last letter?
10. Can you name four organisms beginning with the letter L?
11. How many organisms were there that begin with the letter H?
12. Can you name five organisms beginning with the letter S?

My How You've Changed!

Through the course of years our body takes on a few changes. None more dramatically as the change from the fetal stages to the adult stages. This exercise will allow you to compare a "typical" fetus's body proportion, to the author's body proportion and your body proportion. Some other pertinent information will also be collected.

First, have someone help you measure the following body parts (you may use the English or Metric rule):

_____ 1. The distance from the top of the head to the bottom of the chin

_____ 2. The distance from the bottom of the chin to the shoulder (neck)

_____ 3. The distance from the shoulder to the tip of the fingers

_____ 4. The distance from the shoulder to the waist

_____ 5. The distance from the waist to the feet

_____ 6. The distance from the shoulder to the feet

Second, divide each of the above values by the total length of the body (height). This will give you the relative body proportions. Lastly, multiply each of the body proportions by one hundred sixty (x160) to obtain a relative scale. Place these values in the chart entitled "Body Proportions." You will note two things about this chart. The author's body proportion is given and the fetal body proportions are not given. In order to complete the chart, the measurement of the fetus is necessary. Use the diagram of the fetus given on the following page. Again, use the same procedure given above for measuring the fetus and enter the data.

The last phase of this exercise can now be completed. With your known body proportions, diagram, as well as you can, an outline of your body (similar to that of the fetus and the author). Having completed both phases of the exercise you can see the differences in growth both physically and mathematically!

BODY PROPORTIONS

Subject / Body Parts	Fetus	Author	You
Head (To Chin)		19.1	
Neck (To Shoulder)		9.6	
Shoulder (To Finger Tip)		64.5	
Shoulder (To Waist)		43.0	
Waist (To Feet)		86.0	
Shoulder (To Feet)		129.0	

SCALE: 20 units/division

FETUS --- 8 WEEKS

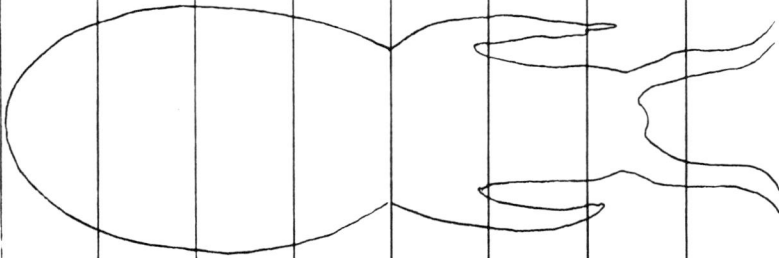

Weight: less than 1 oz.
(or 0.03 Kgm)

HEIGHT: Approximately 1.5 inches
(or 3.8 centimeters or 0.038 meters)

AUTHOR --- 45+ years

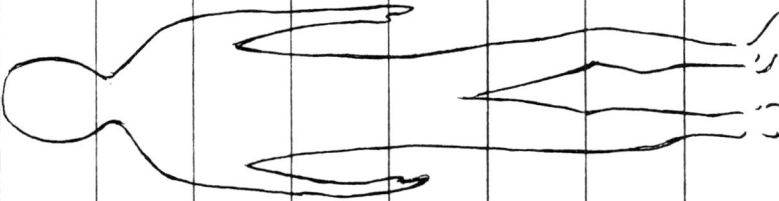

Weight: 140 pounds
(or 63.63 Kgm)

HEIGHT: 67 inches
(or 1.70 meters)

YOU ---- _____ years

Weight: _____ pounds
(or _____ Kgm)

HEIGHT: _____ inches (or _____ meters)

137

Chip off the Old Block*

We are made up of many characteristics (phenotypes) giving rise to many different possible combinations. Below are six characteristics of which there are 64 different combinations. Which one (number) are you? In order for you to determine which combination you are start with the middle of the "block" and work towards the edge. For instance, are you able to roll your tongue? If so, shade in the area with the capital "T". If not, shade in the two small "t's." (NOTE: If you shaded in the large T your combination number could be any number between 1–24 and 57–64.) Now continue with the next series of characteristics given below until each area is shaded (see example below). Eventually, you will arrive at your unique number.

Characteristics—Phenotype

T__	Tongue Rolling	tt	Non-tongue Rolling
E__	Dark Eyes	ee	Light Eyes
A__	Ear Lobes	aa	Attached Ear Lobes
F__	Bent Finger	ff	Straight Finger (little)
L__	Right Handedness	ll	Left Handedness
D__	Dimples	dd	Non-dimples

Reprinted with permission from Chow: Human Biology Lab Manual: A Matter of Survival. Copyright 1987 Kendall/Hunt Publishing Company.

Now compare your number with members of your family to determine the similarities and differences.

Dad: #_____   Mom:#_____   Brother(s):#_____

Sister(s): # _____   Other: # _____

Compare your number with non-members of your family to determine the similarities and differences.

Author of text: # 14 Other: Name/no. _____

Other: Name/no. _____   Other: Name/no. _____

Other: Name/no. _____   Other: Name/no. _____

Things to Ponder

1. Characteristics with capital letters represent dominant traits while the lower case letters represent recessive traits. Were your traits primarily dominant or recessive? Your immediate family? Non-members of the family

2. Conservatively, if we are made up of 1500 different characteristics, how many different combinations are possible?

3. Are you really a chip off the old block? Explain your answer.

4. In light of the exercise completed can you explain why very few people look alike?

Are You a Survivor?

You have taken off from Earth and your space crew is in good spirits as you head for the planet Nucleus.* Happily, you have been informed that there is already a space headquarters set up by friendly aliens who await your arrival on the lighted side of Nucleus. As you approach the planet your spaceship experiences some mechanical difficulties and you are forced to land some 400 Km away from the meeting point with the friendly aliens. Some of crew members are hurt; you must attend to their needs; also, time is of the essence for you to reach the meeting point (the aliens may not remain too friendly).

Some of the survival items you brought along on the trip were destroyed and since survival depends on you reaching the meeting point you must bring only those items which are deemed important. Below is a list of the items. Your task is to rank each of the items in order of their importance, number 1 being the most important, number 2 being the second most important and so on until all items have been ranked.

_____ Two .45 calibre pistols

_____ Life raft

_____ Parachute silk

_____ Food concentrate

_____ Box of matches

_____ 5 gallons of water

_____ Magnetic compass

_____ Two 100-lb. tanks of oxygen

_____ 50 Feet of nylon rope

_____ Stellar Map (Nucleus's constellation)

_____ First aid kit containing injection needles

_____ Portable heating units

_____ Solar-powered FM receiver-transmitter

_____ One case of dehydrated Pet milk

_____ Signal flares

*Similar to atmosphere, terrain, and the environment of Earth's moon.

Evaluation

Compare your ranking of the items with answers given in the book and find the absolute difference between the two rankings. For example if you ranked the Magnetic compass as "5" and the answer is "10," the absolute difference would be "5." This would be the point value for getting the answer wrong. Now take the sum of the differences of all of the items and grade yourself using the following criteria:

Grading System

0–20 points	____	You're a survivor
21–30 points	____	Well, you've got a slight chance of surviving
31–40 points	____	Better take along a guide
41 and over	____	Stay home and study some more biology

Items	Your Ranking	Answers	Difference
Two .45 calibre pistols	_____	_____	_____
Life raft	_____	_____	_____
Parachute silk	_____	_____	_____
Food concentrate	_____	_____	_____
Box of matches	_____	_____	_____
5 gallons of water	_____	_____	_____
Magnetic compass	_____	_____	_____
Two 100-lb. tanks of oxygen	_____	_____	_____
50 Feet nylon rope	_____	_____	_____
Stellar Map	_____	_____	_____
First aid kit	_____	_____	_____
Portable heating units	_____	_____	_____
Solar-powered FM receiver-transmitter	_____	_____	_____
One case of dehydrated Pet milk	_____	_____	_____
Signal flares	_____	_____	_____

I've Got Rhythm . . . Biorhythm That Is . . .

Here's a procedure that you can use to predict your future with regards to your physical being, emotional stability, and your intellectual capacity. Perhaps this is the technique you've been looking for to determine whether or not it's the right time to gamble, to get married, to invest in the business venture, etc. It's easy to do, doesn't require a lot of time, and will be usable for the rest of your life. Just follow the simple procedures and explanations below.

1. Write down today's date (month, day and year) _____

2. Multiply your present age by 365 _____

3. Divide your age by four (omit the decimal point and do not round off to the nearest whole number) _____

4. Calculate the number of days old you were from your last birthday to today's date (you may wish to use the information below) _____

month	days	month	days
1	31	7	31
2	28	8	31
3	31	9	30
4	30	10	31
5	31	11	30
6	30	12	31

5. Determine the sum of sections 2, 3, and 4 _____

6. Divide the sum (in section 5) by the following numbers: 23, 28 and 33. Do the calculation by long hand to obtain the remainder (whole number).

7. Record the remainders for each of the divisors below:

   (a) 23 (physical cycle) _____

   (b) 28 (emotional cycle) _____

   (c) 33 (intellectual cycle) _____

8. The remainders in section 7 are to be used to determine at what point of the curve of the physical, emotional, and intellectual cycle you were at as of today's date. Looking at the physical curve you will note the vertical lines; these lines represent the number of days for a given cycle (1 line = 1 day). The number of days for a complete physical cycle is 23 days after which the curve pattern is repeated. Using your remainder for the physical curve calculations locate the number of days on your curve. This point represents today's date (see example).

Biorhythm Curves

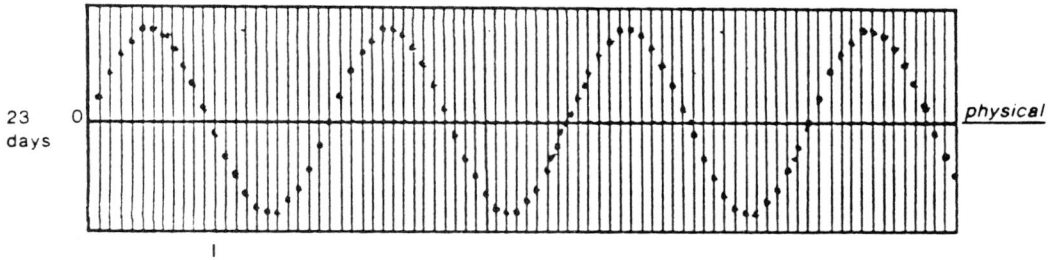

23 days   0   *physical*

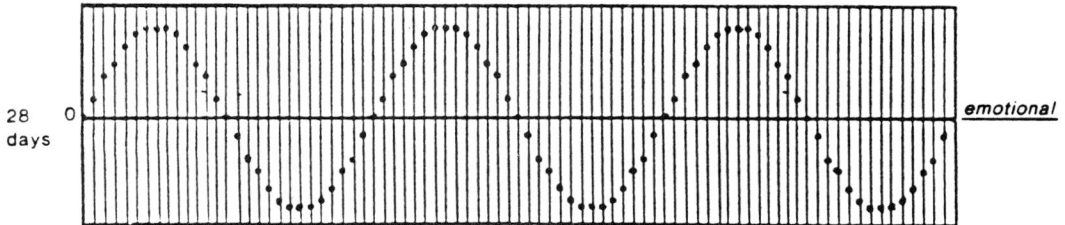

28 days   0   *emotional*

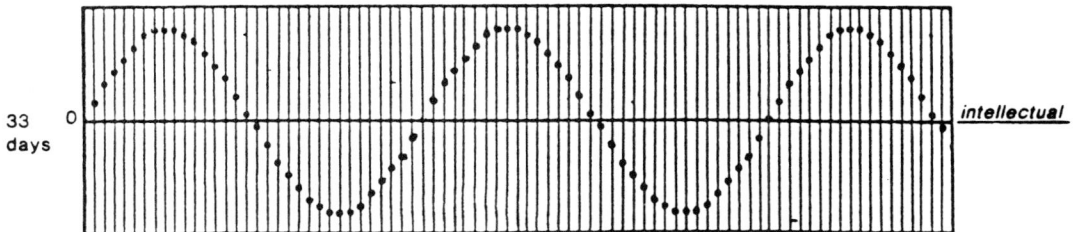

33 days   0   *intellectual*

143

9. Below you will find a time line that somewhat resembles the curves on the preceding page. Match the vertical and horizontal lines from the physical curve onto the vertical and horizontal lines of the time line. (This may be done by holding the papers in the air, against an opaque projector, etc.). Trace the curve onto the time line. Now repeat this process for the emotional and intellectual curves. (Several time lines are given below—in the event of errors!) Label each curve using different colors, different type of lines, etc.

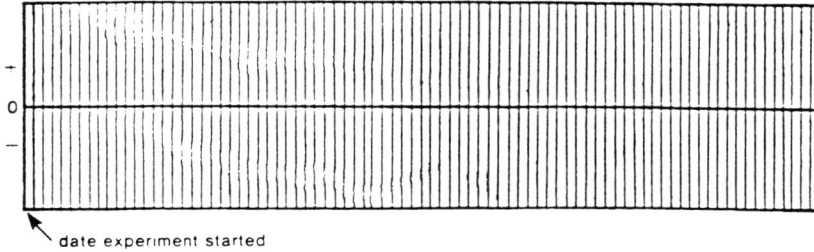

date experiment started

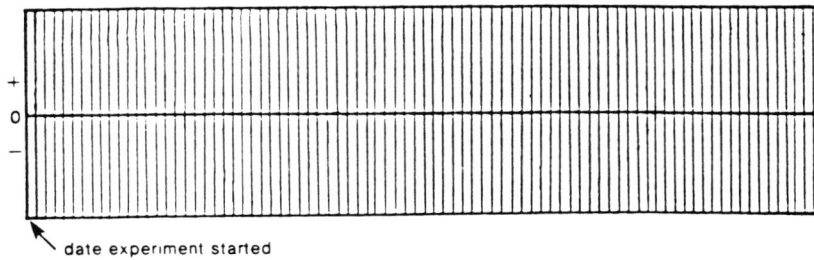

date experiment started

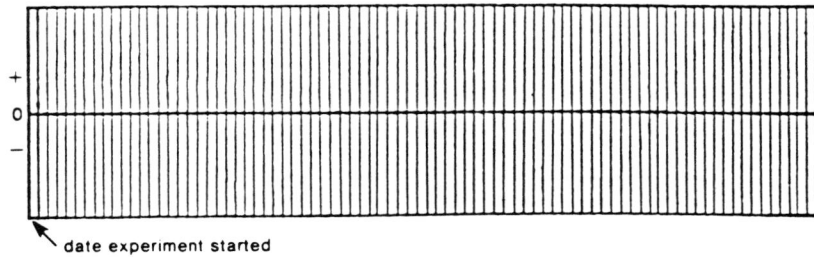

date experiment started

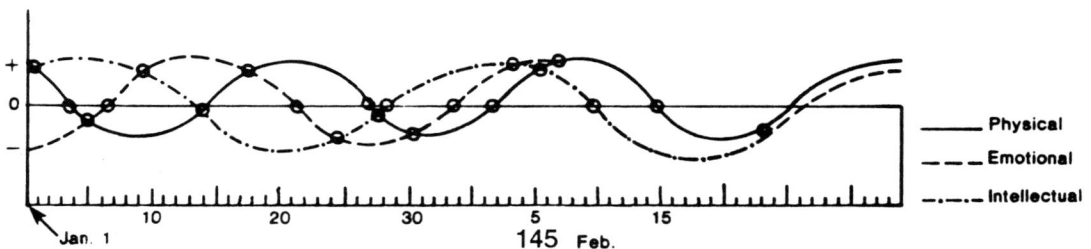

—————— Physical

— — — Emotional

—.—.— Intellectual

Jan. 1   10   20   30   5   15

145 Feb.

10. Interpretation of curves—In the example on the previous page you will note that there are many circled areas. These are referred to as critical days. The critical days may be good (+), bad (–), or neutral (0). What determines a critical day is when a curve crosses the horizontal line, when curves cross each other, or both of the above. The results can then be tabulated in a table format as shown below. Important to note is that biorhythm predictions are given on critical days (that is, not every day is critical! Some days are skipped. Also, on critical days not all three events may be predicted (see Jan. 1; N.P. = no prediction). Now see if you can fill in the Table for Jan. 7, 9, & 25.

SUMMARY OF PREDICTED EVENTS

Date	Predicted Events		
	Physical	Emotional	Intellectual
Jan. 1	+	N.P.	+
Jan. 4	0	N.P.	N.P.
Jan. 5	–	–	N.P.
Jan. 7			
Jan. 9			
Jan. 25			

Answers: Jan. 7 . . . P(N.P.); E(0); I(N.P.)
Jan. 9 . . . P(N.P.); E(+); I(+)
Jan. 25 . . . P(N.P.); E(–); I(–)

Now read your own biorhythm curves for the future. Enjoy!

Biological Dominos

Here's an old fashioned game for two players who are familiar with the organ systems of the human body. It's called Biological Dominos. Instead of matching the number of circles, one merely matches the organs, words, or terms related to the organ system. For example, if a tile contains the words "blood" and "bronchus," a matching tile could be heart for the "blood" (circulatory system) or larynx for the "bronchus" (respiratory system). It is highly recommended that the biological tiles be glued onto a cardboard surface before usage.

Here are the rules of the game:

1. Flip a coin to determine who goes first
2. Place all of the biological domino tiles into a paper bag (non-transparent). Shake the bag to mix the tiles.
3. Each person is to draw out a total of five tiles from the bag
4. Tiles are to be held in each player's hands
5. The first player places a tile down
6. The second player tries to match the tile
   (a) if the tile is a match, the first player will take his turn
   (b) if the tile is not a match, the second player must draw an additional tile from the bag before the first player takes his turn
   (c) the above process is repeated for both players thereafter
7. The game continues until:
   (a) one player has used up all of his tiles; he is declared the winner of the game
   (b) all the tiles have been removed from the bag and neither player can match the tiles; the person with the fewest number of tiles is declared the winner of the game

Unfamiliar with the organs, words, or terms in the various organ system? Here is a checklist for you to learn or review. Have fun with Biological Dominos!

Organ Systems

Digestion: stomach, teeth, esophagus, spleen, pancreas, tongue, small intestines, large intestines

Circulatory: aorta, ventricle, atrium, blood, vein capillary, artery, plasma

Respiratory: turbinates, olfactory bulb, larynx, vocal cord, trachea, bronchus, bronchioles, alveolar sacs

Reproduction: ovary, testes, vas deferens, Fallopian tube, clitoris, penis, vagina, prostate

Endocrine: hormone, pituitary, thyroid, adrenals, first messenger, pineal body, thymus, parathyroids

Nervous: neuron, cerebrum, meninges, cerebellum, spinal cord, pons, medulla, axon

Skeletal: bone, cartilage, radius, ulna suture, synovial joint, atlas, femur

STOMACH	TEETH	ESOPHAGUS	AORTA	SPLEEN	TURBINATES	PANCREAS	OVARY
TONGUE	HORMONE	SMALL INTESTINES	NEURON	LARGE INTESTINES	BONE	VENTRICLE	ATRIUM
BLOOD	OLFACTORY BULB	VEIN	TESTES	CAPILLARY	PITUITARY	ARTERY	CEREBRUM
PLASMA	CARTILAGE	LARYNX	VOCAL CORD	TRACHEA	VAS DEFERENS	BRONCHUS	THYROID

BRONCHIOLES	MENINGES	ALVEOLAR SACS	RADIUS	FALLOPIAN TUBE	CLITORIS	PENIS	ADRENALS

VAGINA	CEREBELLUM	PROSTATE	ULNA	FIRST MESSENGER	PINEAL BODY	THYMUS	SPINAL CORD

PARATHYROIDS	SUTURE	PONS	MEDULLA	AXON	SYNOVIAL JOINT	ATLAS	FEMUR

153

A Game of Cross Out

Instructions: Taking turns with your opponent your task is to cross out words that have a common letter. The first person to have three words sharing a common letter wins. Your opponent will try to stop you by crossing out one of your words. Use different colors (pen or pencil) to keep track of your words. Four sets of games are provided for you.

Game #1	Game #2	Game #3	Game #4
Funny	Funny	Funny	Funny
Loop	Loop	Loop	Loop
Fat-cup	Fat-cup	Fat-cup	Fat-cup
Tick	Tick	Tick	Tick
Life	Life	Life	Life
Pen	Pen	Pen	Pen
Cherry	Cherry	Cherry	Cherry
Iron	Iron	Iron	Iron
Sly	Sly	Sly	Sly

Question: Which word has letters that are common to all of the other words?

PART NINE

Quickies

The activities here do not have anything to do with the title of this section. They require time, creativity, imagination, and some biological background. Some of the activities are the hardest ones I could come up with to challenge you to hours of sleepless nights. It will be difficult for you to "not look" at the answers provided in the back of the book. Nevertheless, enjoy your moments and revisit this section as often as you wish. Perhaps, you may wish to give your friends some sleepless nights too! (Oh yes, some of the activities presented here are easy too; I don't want to discourage you—the last puzzle in this section is definitely for you. . . .)

Help!

Two confused hospital attendants were in charge of the maternity ward and gave some of the babies to the wrong mothers. Two situations are shown below where attendant #1 correctly paired two of the babies and attendant #2 correctly paired three of the babies. You have just been hired to be attendant #3. With the given information please pair the babies to their mothers correctly!

Attendant #1

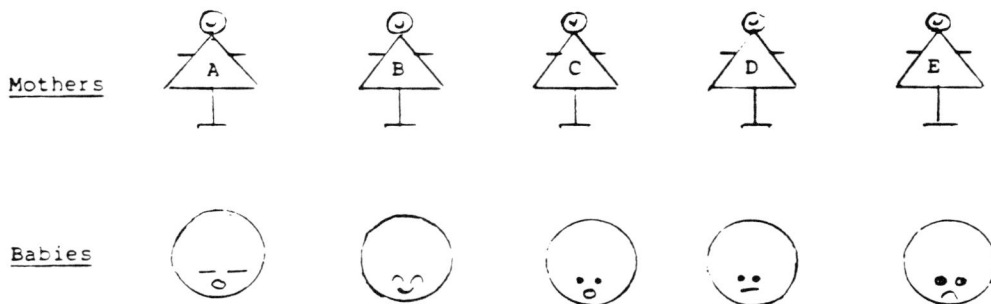

Attendant #2

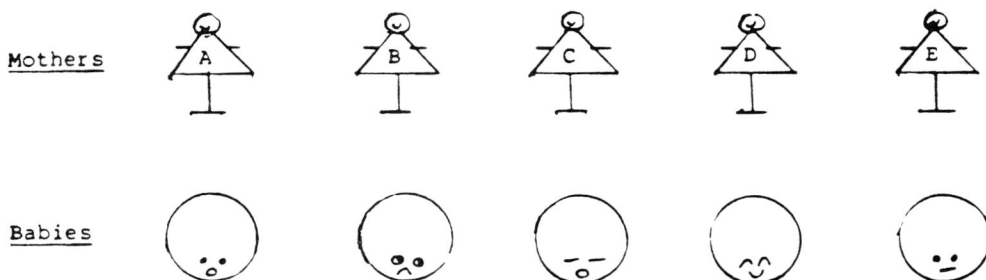

Biological Blanks for the Creative Biologist

Instructions: Each equation below contains the initials of words that will make it correct. Find the missing words. For example: 7 = T of S would be 7 = Types of smell.

1.   7 = T of S   Types of smell _____

2.   12 = P of C N _____

3.   5 = D B in N A _____

4.   4 = C in the H H _____

5.   120 = S P in B P _____

6.   22 = C A A _____

7.   4 = M F G _____

8.   8 = E A A _____

9.   2.2 = P in O K _____

10.   20 = P of A which is O _____

11. 1000 = G in a K _____

12.   72 = H B P M _____

13.   6 = M E in M _____

14.   37 = D C in the H B _____

15. 300–500 = M S produced by M _____

16.   8 = W of E D _____

17.   46 = C in the H C _____

18.   9 = K in O G F _____

19.   5 = T of L in B _____

20.   10 = C F in the M S _____

21. 206 = B in the H B _____

22.   3 = T of N in the N S _____

23.   4 = T of T _____

Egg Hunt

How many eggs (spelling) can you find hidden in this puzzle? Work horizontally, vertically or diagonally but always in a straight line.

```
            G  E  E
         G  E  G  G  E
      E  E  G  G  E  G  G
   E  G  G  G  E  E  E  G  G
   G  G  E  E  G  G  G  E  E
   G  E  G  G  E  G  E  G  G
   E  G  E  E  G  G  E  E  G
   G  E  E  G  G  E  G  G  G
   G  G  G  E  G  G  G  E  E
   E  G  G  E  G  G  E  G  G
   E  G  E  G  E  G  G  E  G
      E  G  G  G  E  G  E
         G  E  E  G  G
```

Now, how many eggs (spelling forwards and backwards—egg or gge) can you find hidden in this puzzle? Work horizontally, vertically or diagonally but always in a straight line.

Numbers, Numbers, Numbers

How much food does it take to feed a person in an average lifetime? Match the numbers with the items.

numbers	items
80,000	beef (tons)
4	potatoes (tons)
1/2	fresh vegetables (tons)
108,000	fresh fruit (tons)
2,900	chicken (tons)
2,000	fish (tons)
4	eggs (number)
4	sugar (tons)
1,800	cheese (tons)
1/2	bread (slices)
20,000	soda (gallons)
$3^1/_2$	milk (gallons)
880	tea (gallons)
296	wine (gallons)
3	coffee (cups)

A Cell and Clock Puzzle

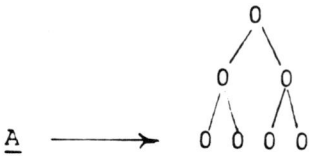

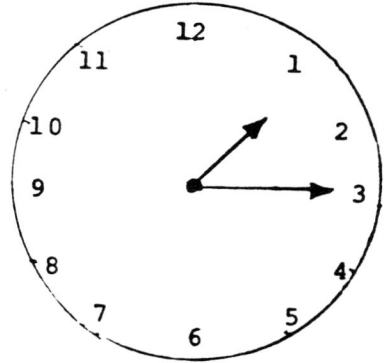

A certain cell is able to reproduce every 20 minutes. Starting at stage *A* of the divisional process and if one hour and forty minutes elapses, what time will the clock read and how many cells will be produced?

Number Please

Instructions: The following telephone numbers represent commonly used words in biology. Can you figure them out? Hint: Look at your telephone.

1. 264-6257

2. 246-5649

3. 326-5649

4. 676-6747

5. 786-6224

6. 368-7679

7. 353-6368

8. 682-5387

9. 822-8653

10. 776-8346

11. 233-6463

12. 752-7643

13. 634-6747

14. 234-7673

15. 796-2773

16. 224-6842

17. 682-7437

18. 255-3537

19. 782-3789

20. 797-8653

21. 437-2746

22. 637-4766

23. 623-7489

24. 456-8847

25. 467-8546

26. 268-4436

27. 872-2432

28. 849-7643

29. 874-2377

30. (638) 226-5476

Gametes Galore

Instructions: Four pairs of gametes resulted from one mother cell (parent cell). Pair up the gametes to one another and to their respective parent cell given information about each gamete below. Only two gametes may be paired to their respective parent cell.

Easy? Now determine which gametes each of the parent cells could not form as a result of (1) a single cross or (2) a double cross.

Half and Half

Familiar words of the Nervous System are shown below but have been cut in half. Recombine them to spell the words.

ORAL

SPIN

CERE

FRON

CRAN

OCCI

DEN

DULLA

AX

TAL

IAL

MUS

DRITE

GES

TROY

TER

BRUM

AL

ON

TEMP

MED

MAT

RON

PITAL

MENIN

NEU

OLAFC

THALA

Can You Measure Up?

A lab technician was presented with a problem. It appears that all of his graduated cylinders were being used except for a very old and forgotten one that had only three markings on it. (See illustration)

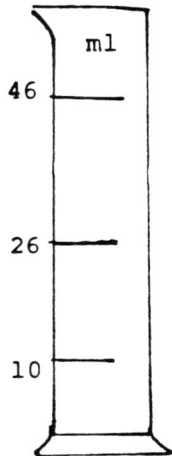

Solve the following problems:

(a) Help the technician measure out 62 ml of water in 3 measures (moves).

(b) Now, repeat the above process but in 2 measures (moves).

(c) What is the fewest measures (moves) required to obtain a volume of 76 ml?

Now Pay Close Attention

Read the sign and without looking say the words aloud or write them down on a piece of paper.

```
        /\
       /  \
      /    \
     / LIFE \
    / IN THE \
   / THE SPRING \
  /_____\
```

Count the number of F's in the following passage below. Go through the passage quickly and do not stop or go back at the beginning.

> FINISHED FILES ARE THE RESULT OF YEARS OF SCIENTIFIC STUDY COMBINED WITH THE EXPERIENCE OF MANY YEARS.

Total number of F's _____

Nutrition

What do these six foods have in common?

BEANS

BREAD

CABBAGE

DOUGH

LETTUCE

PEANUTS

Your Male-Female Index

Instructions: How much do you know about males and females? Below are listed a variety of problems, ailments, diseases, etc. Simply place the letter M for males and the letter F for females beside each of the terms that you feel are more applicable to a particular sex. Check your score below.

_____	Accident	_____	Inguinal hernia
_____	Anemia	_____	Kidney stones
_____	Cancer	_____	Leprosy
_____	Diabetes mellitus	_____	Lupus
_____	Femoral hernia	_____	Migraine
_____	Gallstones	_____	Myasthenia gravis
_____	Gastric ulcers	_____	Obesity
_____	Gout	_____	Osteoporosis
_____	Heart disease	_____	Rheumatoid arthritis
_____	Hepatitis	_____	Substance abuse
_____	Hypertension	_____	Tuberculosis

Scoring

A perfect score = Good for you! You're very knowledgeable.
5 wrong   = You've got the facts!
10 wrong   = You're guessing . . . take a biology course. . . .

Calculating Biology

Numbers are very meaningful. Perform the following calculations with your calculator and discover the world of biology.

The Calculations

1a. 35 x2
  b. x 10
  c. + 10
  d. turn calculator up-side-down

2a. 10,000/2
  b. + 600/2 + 30
  c. + 8
  d. turn calculator up-side-down

3a. 77 x 200
  b. + 69
  c. x 5
  d. turn calculator up-side-down

4a. $9000 - ((16)^{1/2} \times 10^3)$
  b. $+ (30)^2$
  c. $+ (3)^2 - 2$
  d. turn calculator up-side-down

5. Subtract 3 from the previous answer in #4 and turn the calculator up-side-down

6a. 1/5 x 25
  b. $+ 10^2$
  c. $+ (50)^2 + 13,500/3$
  d. turn calculator up-side-down

7a. $(4 \times 10^{12})^{1/2}$
  b. $+ (81 \times 10^4)^{1/2}$
  c. $+ (10^4)^{1/2} + (8100)^{1/2} + (16)^{1/2} - (900)^{1/2}$

8a. $(300)^2 \times 10^2$
  b. $-10^6 + 10^5$
  c. $+ 10^3 + (3^2 \times 10) - (9)^{1/2} \times 10^1 + 2^2$

Clue

1. If you're under pressure this is what would result. . . .

2. Hi honey!

3. You're a slug without it!

4. It's like keeping a diary.

5. Rhymes with #4

6. It's clothes time when you roll in this. . . .

7. You animal!

8. It's what life is all about. . . .

171

White Blood Cells on the Move

Part I

Instructions: Show how 5 neutrophiles, 5 basophils, 5 lymphocytes, 5 monocytes, and 5 eosinophils could occupy each of the cells (squares) below such that no two white blood cells of one kind will be found together vertically, horizontally or diagonally. Use the letter N = neutrophil, B = basophil, etc.

White Blood Cells on the Move

Part II

Instructions: The following answers were given as a possible answer to Part I of this puzzle. Obviously the solution is incorrect (but not by much). However, we can still salvage a bit of pride. Move two of the neutrophils (switch places with any of the other white blood cells) such that no two neutrophils are in the same line vertically, horizontally or diagonally.

E	M	L	N	B
B	E	B	E	L
L	N	E	B	M
M	B	L	M	N
N	M	N	L	E

A State of Confusion (Entropy)

A confused Plantanimal cell was determined to be organized. The cell organized itself into sixteen (16) spaces in which subcellular particles are to be placed. There were four (4) mitochonrida, three (3) ribosomes, three (3) vacuoles, three (3) chloroplasts, and three (3) centrioles.

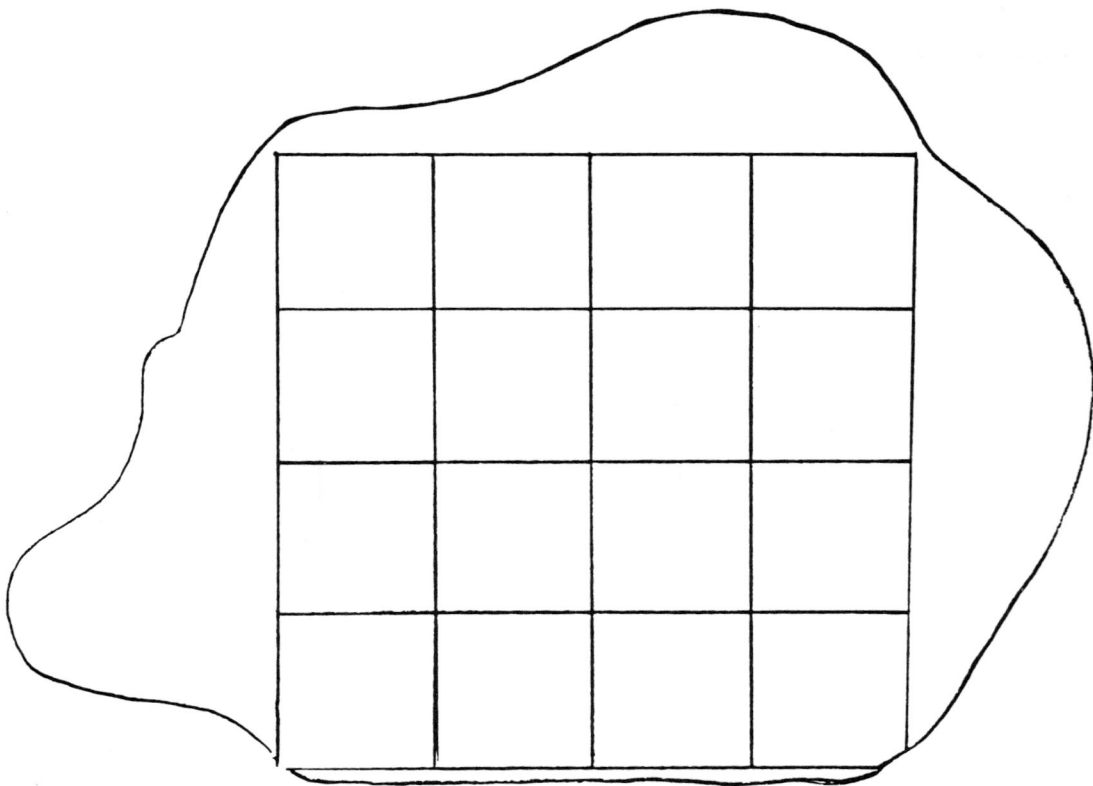

How can the subcellular particles be placed in the spaces (one subcellular particle per space) such that no *two* particles of the same kind are in a line vertically, horizontally or diagonally. Help the Plantanimal cell out (please)!

It's Easy as A,B,C . . .

Question: Why do we have conventional time and military time? Decode the following items below for the answer.

9:20	18:25
9:19	15:21
1:14	20:15
15:20	12:15
8:05	15:11
18:23	1:20
1:25	12:09
6:15	6:05

Here's some timely advice for those who work an eight hour day: Clue: read the title.

18:05	15:05
12:01	14:10
24:01	15:25
14:04	12:09
20:18	6:05
25:20	

I Know What You're Really Interested In. . . .

Instructions: Pick a color of your choice. The color is related to a number within a circle. Move in a clockwise direction the number of spaces as indicated by your color. Move *inward* towards a new encircled number. again, move in a clockwise direction the number of spaces as indicated. Move inward towards a letter. Check the letter below to determine your interest . . . and your interest is . . . (see answer sheet).

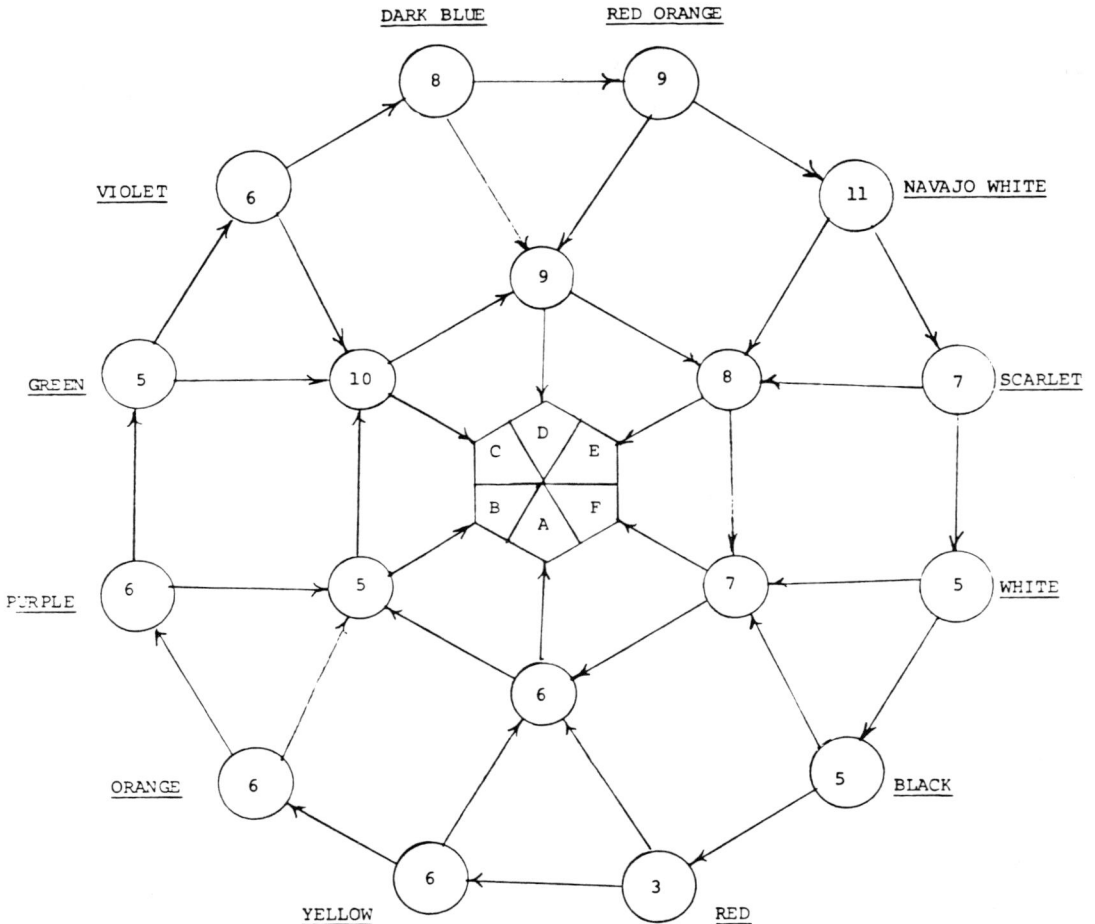

Spectrum of Interests

A = Biology	C = Philosophy	E = Politics
B = Music	D = Education	F = Romance

PART TEN

Answer Sheet

Page 3   Welcome to Biology!

Page 5

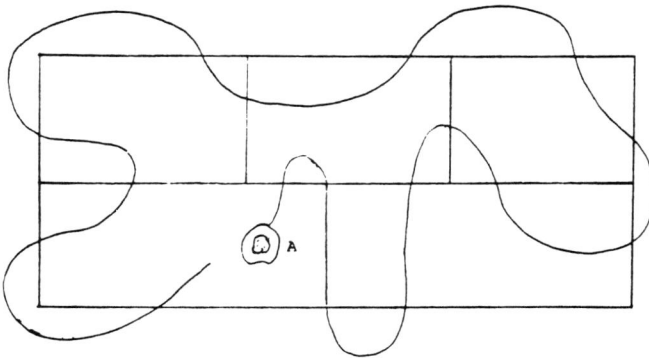

Page 6   Heart, Nose, Brain, Kidney, Stomach

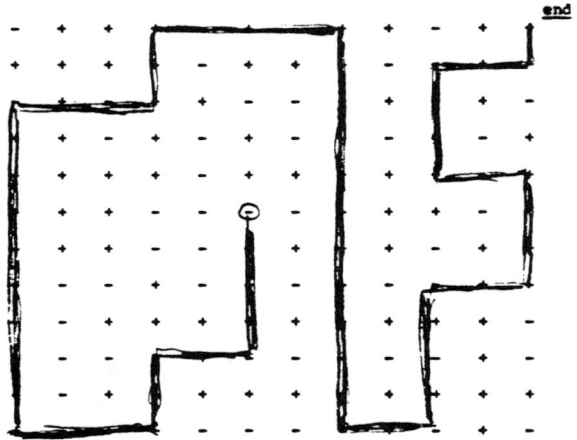

Page 7

Page 8   Equal amounts of calories can be burned during intense concentration or heavy exercise!

Page 9

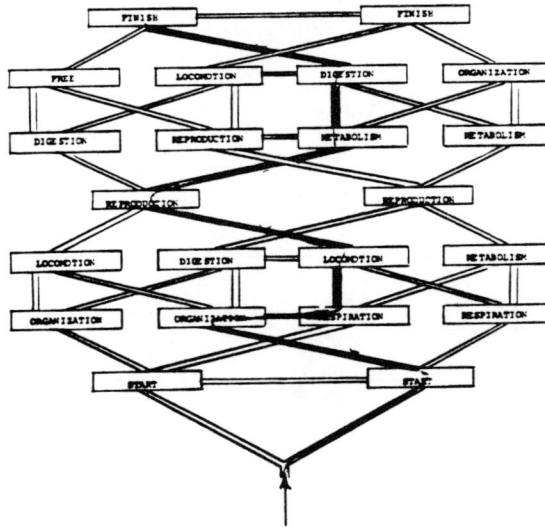

178

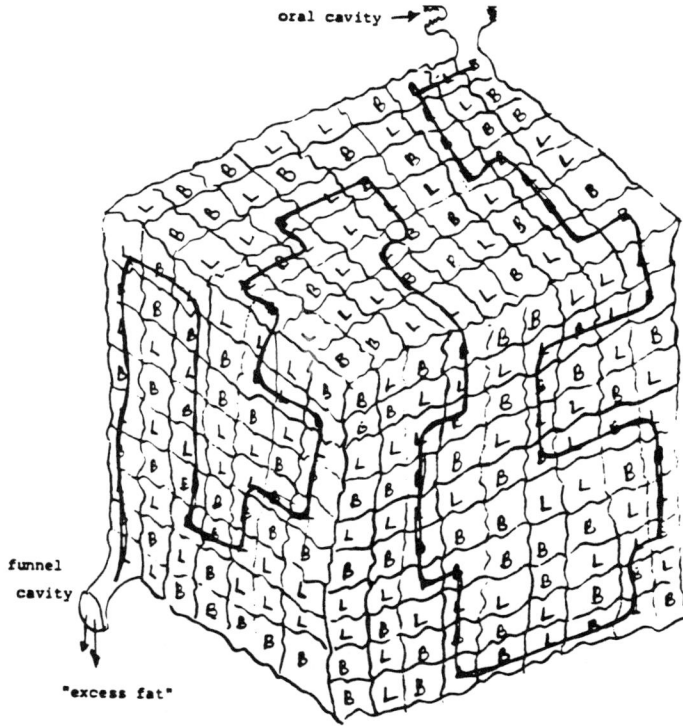

oral cavity →

funnel cavity

"excess fat"

Page 10

Page 11 Muscle-Connective; Epithelial-Blood; Nervous-Bone Brown-Blue; Yellow-Orange; Red-?

Page 12 He's evolving a little slow . . .
Too much starch on the right collar!
Your daughter looks just like you!
How's the macrame class coming along?

Page 13 Those are darling plasmids you're wearing.
I think you need a haircut.
He's wearing his new designer jeans.
I should have permed both sides.

Page 14 Careful girls! There's something strange about him.
Poor John. He hasn't been the same since his better half left him.
He's one of those health freaks!
I understand she's a changed person since she stopped smoking.

Page 15 DNA C: TTTTCCCCGGGGGGAAAAAA
AAAAGGGGCCCCCCTTTTTT

179

Page 15

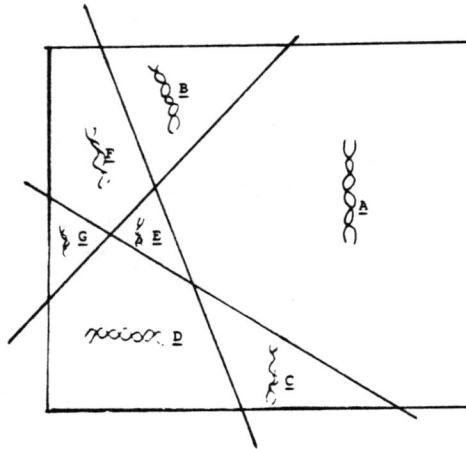

Page 16

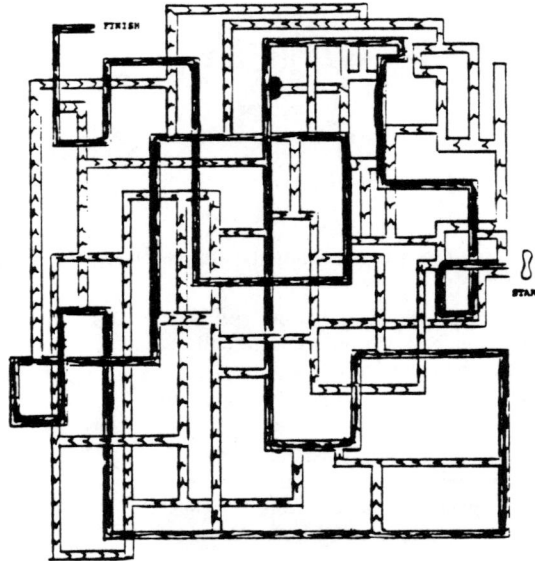

Page 19

1. Herbaceous	2. Antiseptic	3. Chlorophyll	4. Histology
5. Hemoglobin	6. Protozoan	7. Cardiogram	8. Centimeter
9. Microscope	10. Epiderm	11. Anthropology	12. Protoplasm
13. Semipermeable	14. Biology	15. Diffusion	

Page 20

Stiperithecium, Columnareolar, Acneuron, Stigmating Type, Meristembyro, Dialysisotonic, Chromatinterphase, Proteinorganic, Ribosomeiosis, Enzymegavitamines, Jaundicecum, Rumendoparasite, Stomachyme, Pacemakerythrocytes, Flowerosion, Starchloroplasts, Plasmacrophages, Oxyhemoglobintercostals, Aldosteronephron, Urinephron, Antigendotoxin, Generationtogeny, Phenotyperfect flower, Phylumbel, Asexualternation of generation, Capsulenticles, Isogamycology, Cuticleaf, Epiphytendril, Fungicidermatophytes, Chlorosisogamy

180

Page 21   Heavy drinking men are likely to sire girls by ten folds!

Page 22
1. *Heart* Colon Spleen Stomach (digestive versus circulatory system)
2. Carbon Ion *Glucose* Oxygen (atoms versus molecules)
3. Corn Peas Cereal *Spinach* (seed versus leaf)
4. *Sodium* Chloride Fluoride Iodide (anions versus cations)
5. Algae *Amoeba* Fungi Moss (plant versus animal)
6. Cow Sheep *Tiger* Deer (herbivore versus carnivore)
7. *Tapeworm* Tick Leech Flea (ectoparasite versus endoparasite)
8. Cell wall *Centriole* Chloroplast Chlorophyll (plant cell versus animal cell)
9. Uric acid Ammonia Urea *Protein* (waste products versus useful products)
10. Eel Fish Shark *Anemone* (fast moving versus slow moving)

Page 23

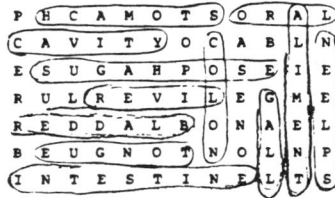

Page 24   Another word for sinoatrial node is pacemaker.

Page 25   The correct sequence of columns are: 4,5,7,1,2,8,3,6 and the hidden words include the following: kidneys, bladder, skin, sweat, urea, ammonia, calyx, uric acid, liver, air, water, oxygen, and ureter.

Page 26

1. Protozoan	2. Chloroplast	3. Mitochondria	4. Metabolism
5. Chromosomes	6. Vertebrates	7. Pacemaker	8. Ectoparasite
9. Gymnosperms	10. Hypothalamus	11. Oxidation	12. Finish

Page 27

1. Biologist	2. Tissue	3. Carbohydrate	4. Chemistry
5. Electrons	6. Lipids	7. Molecules	8. Anticodon
9. Operon	10. Centriole	11. Chromatin	12. Microtubules
13. Ribosomes	14. Permeable	15. Microtome	16. Diffusion
17. Carbuncle	18. Deciduous	19. Epidermis	20. Digestion

Page 28

1. Autotroph	2. Hindbrain	3. Olfaction	4. Thyroxine
5. Cartilage	6. Osteocyte	7. Gallstone	8. Edentates
9. Ingestion	10. Polycythemia	11. Immunization	12. Agranulocyte
13. Tuberculosis	14. Osteoporosis	15. Capillary	16. Emphysema
17. Endoparasite	18. Monosaccharides	19. Vertebrae	20. Atherosclerosis

Page 29

1. Stamen	2. Runner	3. Phloem	4. Anther
5. Mushrooms	6. Horsetail	7. Liverwort	8. Deciduous
9. Epiphytes	10. Annularia	11. Chestnuts	12. Chlorosis
13. Xerophyte	14. Mesophyll	15. Pericycle	16. Spirogyra
17. Gibberellins	18. Gametophytes	19. Angiospermae	20. Chloroplasts
21. Conidiospore	22. Phycomycetes	23. Thallophytes	24. Megagametophyte
25. Photoperiodisms			

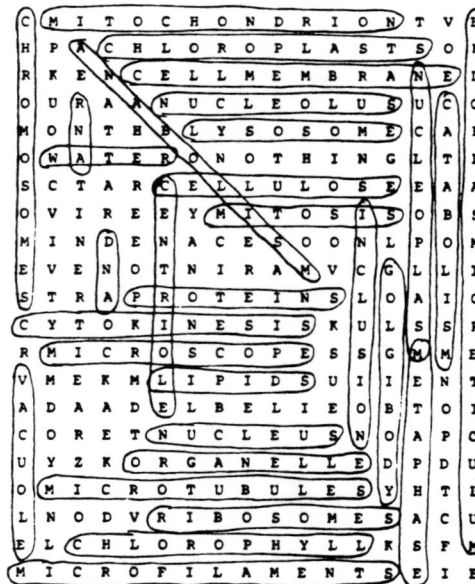

Page 41

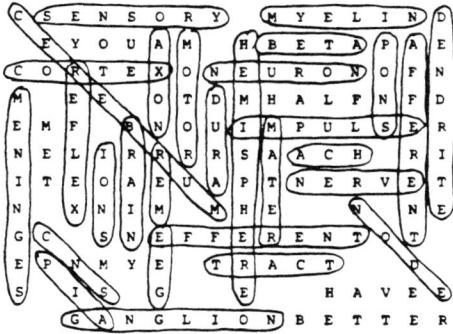

Have you met my better half?

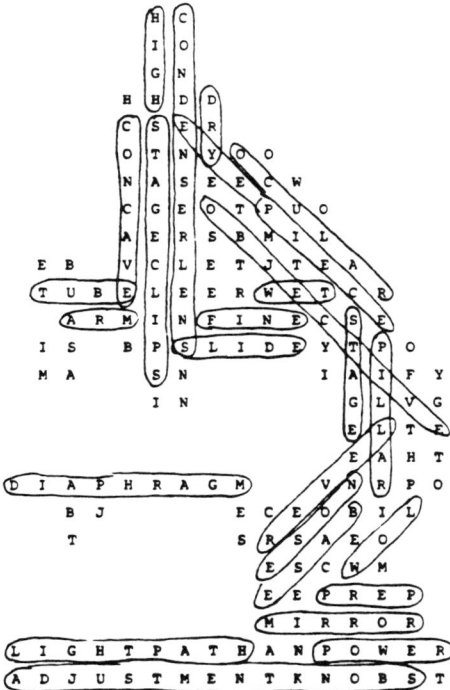

Page 42

183

```
                A I N E R T
              N U C L E U S A A
            G I O N L I P I D
            A T E X B O N D E
            S C T B Y L A L S
            H Y D R O G E N T
              U A T O M E N J
                T S E I O N
                O N E R N D
                L T       E E
                L I S     R C
                  I S     R C

      N G E D N A                  P T U E T S
    O O E S M M W T            O K E N L A I Y
L   U R P U S A T E            T A A E R E E B
I   A B C A T I O N R          I T R N A C D V P
F   R I A E A R N G            S G P R O T O N O
E   K T R D R L O U            Y S A L T R A D K
    E S O T C I P N            E C A R B O N A
    O V V Y H Y                A N I O N N
```

```
                    S E
                P E I N
        S E C N
        E A N T E E
        R H R T E T E
        N A D O H R V
                    W G E
                    D A T
                    U C T
                    A T I
                    E T          E N Z Y M E   C O F A C T O R   G M E T R A
    C Y C L E M                   Y           Z C   G T
  F A S O E L E C T R O N U         Y M A   M R   I
  A D P C Y T O C H R O M E S F A D   E C O   T A   N
  D P G L U C O S E R E D U C T I O N   T B   A N   A
  H A R C A L V I N B E N S O N E N T R O P Y   O B S   D
  R E S P I R A T I O N C A R R I E R S       L O S   H U R
  G L Y C O L Y S I S O X I D A T I O N     I R S   T
  O K R E B S T H E R M O D Y N A M I C S   S T E F
  L           U S I         T S G O O M D M
```

184

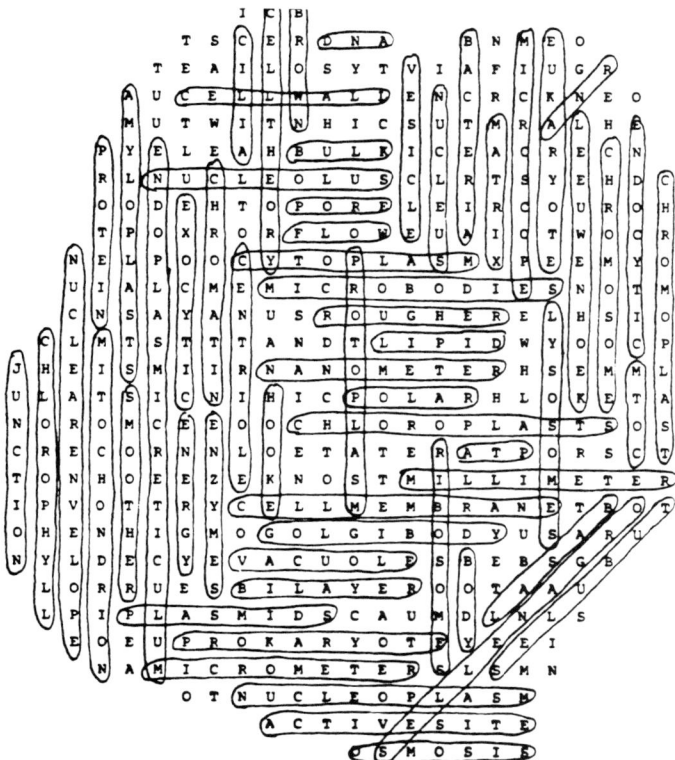

Page 46

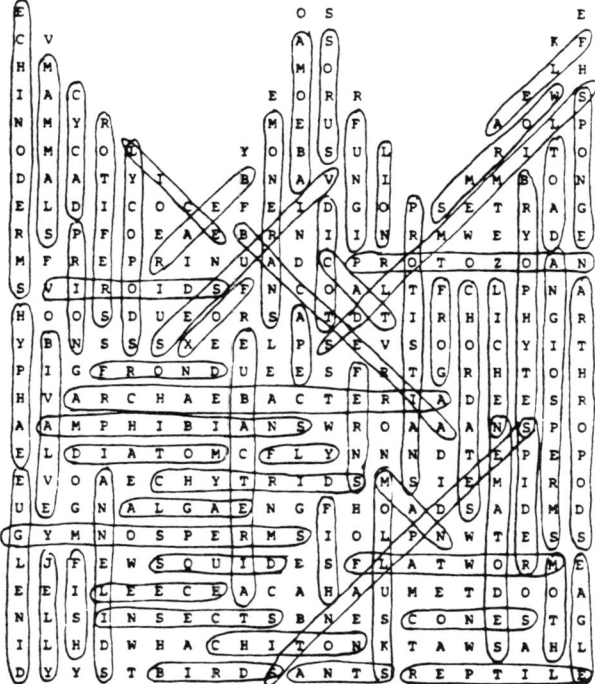

Page 48

185

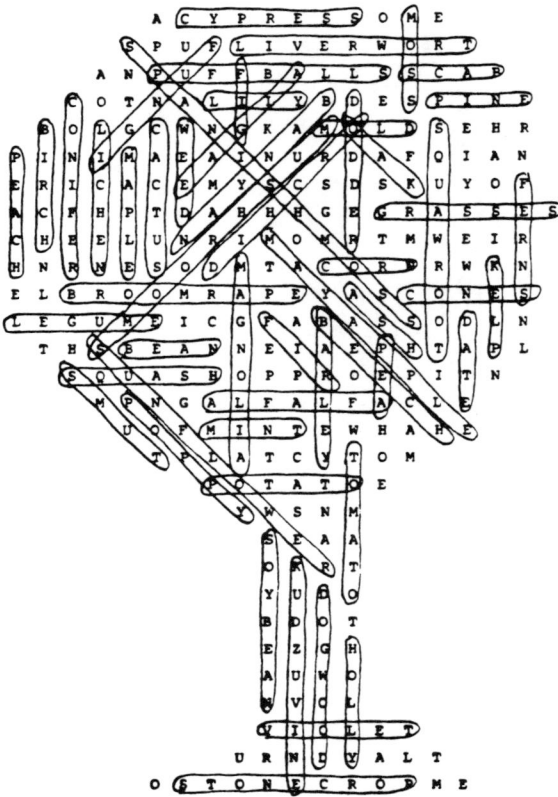

Page 50

Page 52

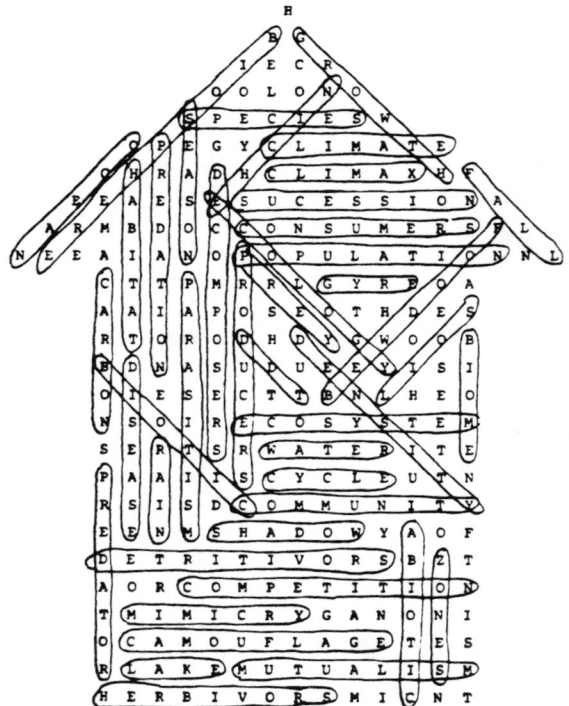

186

Page 54

Page 56

Page 58

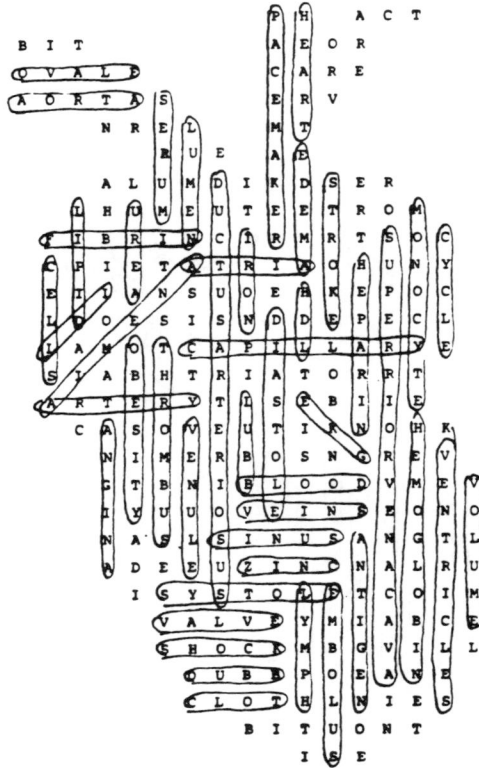

Page 60

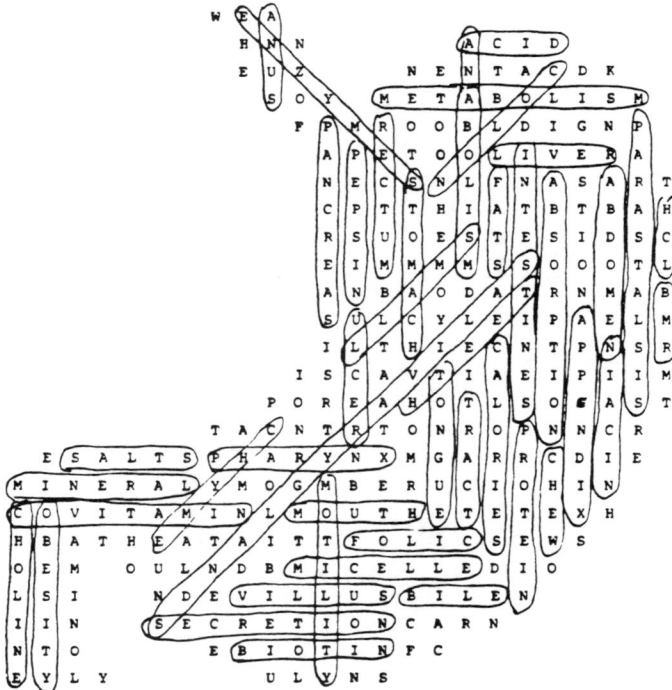

```
                    E P I N E P H R I N E   M E S S E N G E R
                          H Y P O A C T I V E
                          P I T U I T A R Y
                          O       N C E
                          T       H U
                       B  H       I T
                    C     A       B E
                    T     R       I
            T  S M A L P O A   T H Y R O I D      H T
         B  G  T S M A   E  M   I N S U L I N    T E S
       T O  E  H B M L   R  U N I U C O R T I S O L A
       K U  C  R L P I F E G L O R C B L T V I T P
       I O  O  U N N O S N E F E A L O B A L E A H I P
       R I  I  S A E S H A A A S R D T I T H S R L Y M P
       E D  S  L T F O L L I C L E A T I E I G Y E U T
       X N  S  F I R S T I S T I M U L A T I N G  T M H
         S     P R O T E I N H O R M O N E R O H N S U G
       K K S G O I T E R R H Y P E R A C T I V E    L I
         M E D U L L A P A N C R E A S O F T S
             C N T H Y R O X I N E G G S E
             S O M I N A C U T I V E B
```

```
              P N T A D B A S E
              V P E S I N C U S O R
         T I H M L A D E L S U T B D E M
       F U G I E V L O N O K N L O I I A E
       S L A M A T R I X N D C E E T A U D T A K
       I E M I C S A X I A L D I V I S I O N F
       B A R E N E L I M B C T E G O U T V A S S
       I X O T     U I T       U P T H A
       A I N W S     L A N       H Y O F F
       S A S Y Y H I P V H   Q U P B U R S C A
         N V I N C N E T A I   U E R U L V I Y T
         O I A R H O N E D   L B R A C I S T
         V L   S L A E D   O T I T A L K
         I     P I V O T     N W U U D N
         A     R P E V U D O N   R I E A
         L     A E R J O I N T S   E E S
               I F S N B V I C E E S
               N E I E P I P H Y S E S
               Y M A C T S A C R U M P
               O U K H I N G E M S I N
               A R T H R O L O G Y I N
               C V O L A N T H I G H E
               S L I D E K
```

189

Page 66

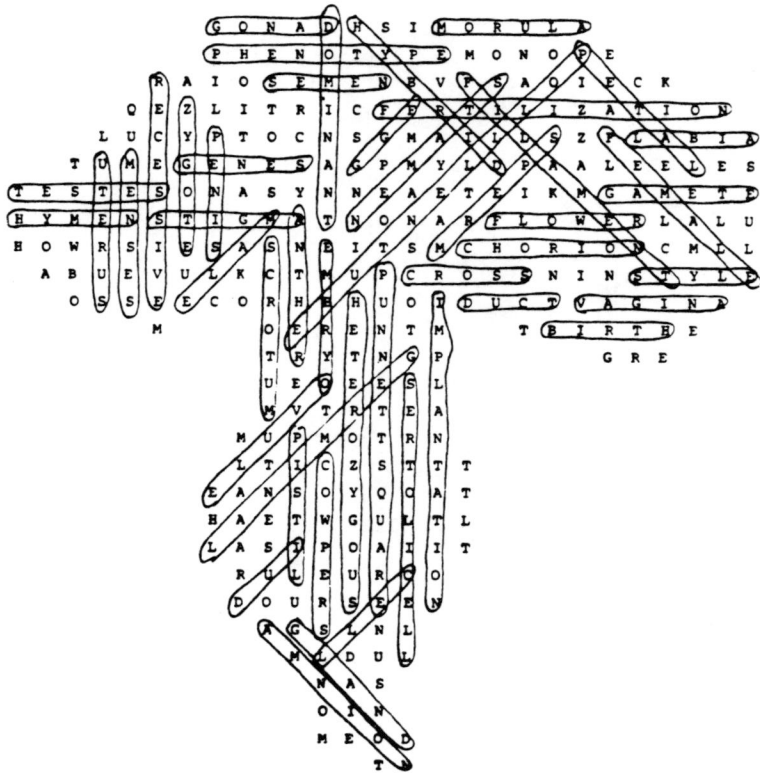

Page 68

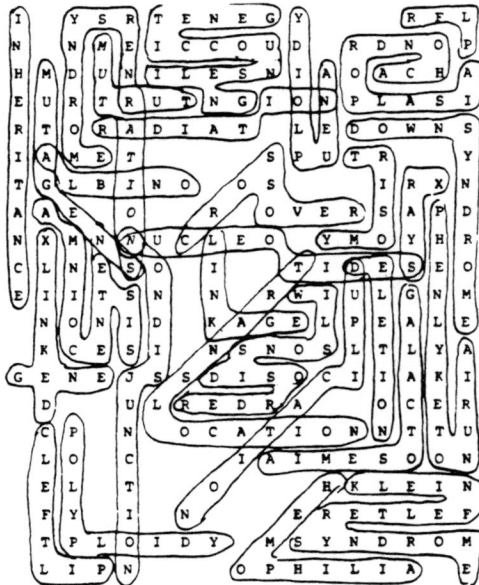

Page 71	The more hair a man has on his chest at age thirty the less hair he'll have on his head at age forty.
Page 72	Each stair you climb adds four seconds to your life according to studies at Johns Hopkins University.
Page 73	According to some studies sixty percent of college students already suffer from high frequency hearing loss.
Page 74	Blood type O men live longer than blood type B men. The opposite is true for women. The muscles of your eyes move one hundred thousand times a day.
Page 75	Pandiculating is a medical term for yawning!
Page 76	Magnolias and water lilies are the first true flower.
Page 77	There are sixty eight people per square mile in the United States.
Page 78	There are about eight children born every minute in the U.S.
Page 79	In DNA base pairing there is an A for every T. There is a G for every C.
Page 80	Seventy five percent of dust floating in your house is made up of dead skin cells.
Page 84	The atom is the smallest portion of an element that still retains the properties of the element.
Page 85	1. NaF 2. Hydrogen 3. Boron 4. Banana 5. 4 x 16 (O) = 64 or 4 x 190 (O's = Os) = 760 6. China (C = carbon, H = hydrogen, I = iodine, Na = sodium) 7. Ice cream 8. NaCl 9. neon/calcium 10. Atomic symbols are all found in carbon. 11. Bring a pound ham 12. 17 13. 9;11 14. Boron; all other atoms have an atomic weight that is twice that of the atomic number 15. W + At + Er − 3(12) = 249 = Berkelium 16. $\underline{KAT}$ + Hot-tin + Ruff (Cat on a Hot Tin Roof) $\overline{A}$

Page 87

Page 88

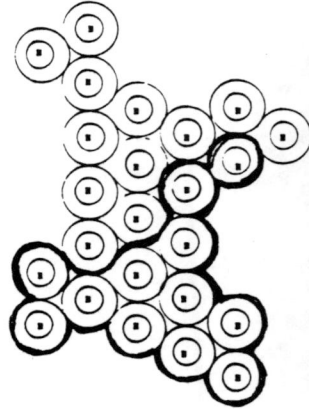

Page 89

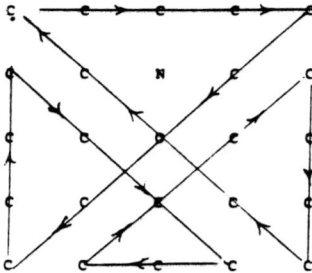

Page 93   Plasma membrane, Plasmodesmata, Chloroplast, Vacuole, Mitochondrion, Ribosome, Endoplasmic reticulum, Golgi apparatus, Nucleus, Nucleolus, Chromatin, Nuclear membrane, Nuclear sap

Page 94   Anther, Filament (Stamen), Stigma, Style, Ovary, Ovules (Pistil), Petal, Sepal, Peduncle, Receptacle

Page 95   Fused pericarp and seed coat, Endosperm, Cotyledon, Coleoptile, Epicotyl, Hypocotyl, Radicle (Embryo), Coleorhiza, Peduncle

Page 96   Blade, Petiole, Stipules, Axil, Internode, Node

192

Page 113

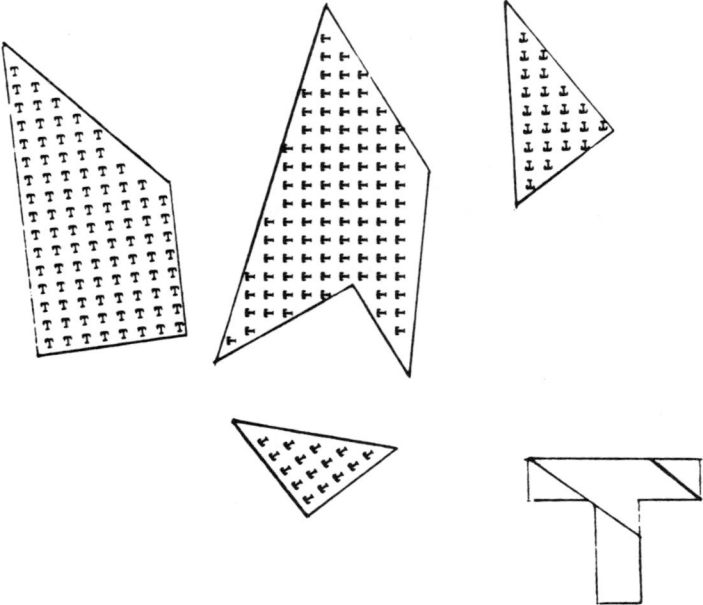

Page 129

194

Page 134	1. Concentration 2. four 3. two 4. four 5. five 6. twenty 7. ten for plants; ten for animals 8. Larch, jellyfish 9. ten pairs 10. Liverwort, larch, lily, lobster 11. Hemlock, horsetail 12. Sponges, snail, shark, squid—there are only four organisms beginning with the letter *s*
Page 141	1. Two 100-lb. tanks of oxygen 2. 5 gallons of water 3. Stellar Map (Nucleus's constellation) 4. Food concentrate 5. Solar-powered FM receiver-transmitter 6. 50 feet of nylon rope 7. First aid kit containing injection needles 8. Parachute silk 9. Life raft 10. Signal flares 11. Two .45 calibre pistols 12. One case of dehydrated Pet milk 13. Portable heating units 14. Magnetic compass 15. Box of matches
Page 155	Life
Page 159	Number the babies one through five from left to right using Attendant #1. The correct match would be as follows: A-1, B-5, C-3, D-2, E-4
Page 160	1. Types of smell 2. Pairs of cranial nerves 3. Different bases in nucleic acid 4. Chambers in the human heart 5. Systolic pressure in blood pressure 6. Common amino acids 7. Major food groups 8. Essential amino acids 9. Pounds in one kilogram 10. Percent of air which is oxygen 11. Grams in a kilogram 12. Heart beats per minute 13. Major elements in man 14. Degree Celsius in the human body 15. Million sperms produced by man 16. Weeks of embryonic development 17. Chromosomes in the human cell 18. Kilocalories in one gram of fat 19. Types of leukocytes in blood 20. Common factor in the Metric System 21. Bones in the human body 22. Types of neurons in the nervous system 23. Types of taste
Page 161	**Fifty-seven (57); One hundred five (105)**
Page 162	Beef (4), Potatoes (4), Fresh vegetables (4), Fresh fruit(3), Chicken (2), Fish (1/2), Eggs (20,000), Sugar ($3^1/2$), Cheese (1/2), Bread (108,000), Soda (2,900), Milk (2,000), Tea (880), Wine (296), Coffee (80,000)
Page 163	Sorry, this is not a real clock. Time will remain unchanged and only four cells were present during that time!
Page 164	1. Animals 2. Biology 3. Ecology 4. Osmosis 5. Stomach 6. Entropy 7. Element 8. Nucleus 9. Vacuole 10. Protein 11. Adenine 12. Plasmid 13. Meiosis 14. Adipose 15. Synapse 16. Abiotic 17. Ovaries 18. Alleles 19. Puberty 20. Systole 21. Heparin 22. Nephron 23. Obesity 24. Glottis 25. Insulin 26. Antigen 27. Trachea 28. Thyroid 29. Triceps 30. Metabolism
Page 165	1-b, f 2-a, g 3-c,e 4-d,h; Any parent cell will result in the formation of 8 different gametes either via pure lines, single crosses, or double crosses. Since 2 conditions were given the only gametes that can not be produced would be the pure lines
Page 166	Neuron, Axon, Frontal, Dendrite, Temporal, Occipital, Cerebrum, Meninges, Thalamus, Spinal, Cranial, Medulla, Matter, Olfactory

Page 167   (a) 26 + 26 + 10 = 62ml
(b) 46 + 16 or 36 + 26
(c) 46 + 20 + 10 = 76ml

first measure: fill to 46ml _____ empty	46ml
second measure: fill to 46ml—empty to 26ml	20ml
from 26ml—empty to 10ml and use the 10ml	10ml
	76ml

Page 168   Life in the the spring; six (6)

Page 169   All of them are in reference to money (slangs)

Page 170   Accidents-M, Anemia-F, Cancer-M, Diabetes mellitus-F, Femoral hernia-F, Gallstones-F, Gastric ulcers-M, Gout-M, Heart disease-M, Hepatitis-M, Hypertension-F, Inguinal hernia-M, Kidney stones-M, Leprosy-M, Lupus-F, Migraine-F, Myasthenia gravis-F, Obesity-F, Rheumatoid arthritis-F, Substance abuse-M, Tuberculosis-M, Osteoporosis-F

Page 171
1. Oil	2. Bees	3. Shell	4. Logs
5. Hogs	6. Soil	7. Zoology	8. Biology

Page 172

N M B E L   N = neutrophil
B E L N M   B = basophil
L N M B E   L = lymphocyte
M B E L N   M = monocyte
E L N M B   E = eosinophil

Page 173

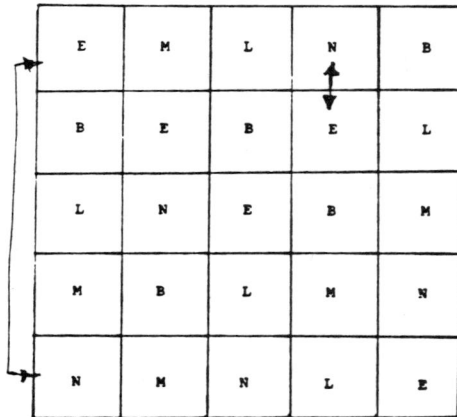

Page 174

V M C Ch   V = vacuole
Ch R V M   C = centrioles
M C Ch R   R = ribosomes
R V M C   M = mitochondria
                  Ch = chloroplasts

Page 175   It is another way for you to look at life.
Relax and try to enjoy life.

Page 176   Biology!!!